# *ARE PESTICIDES REALLY NECESSARY?*

National Council for
**ENVIRONMENTAL
BALANCE, INC.**

4169 Westport Road
P.O. Box 7732
Louisville, Kentucky 40207
502-896-8731

# *ARE PESTICIDES REALLY NECESSARY?*

Published by Regnery Gateway, Inc.
Book Publishers 360 West Superior
Chicago, Illinois 60610

Manufactured in the United States of America

Library of Congress Catalog Card Number: 80-54684
International Standard Book Number: 0-89526-888-4

He who has bread may have many troubles;
He who lacks it has only one.

—Old Byzantine Proverb

# *Table of Contents*

## SECTION III  *Pesticide Safety*

*Chapter*

# Preview

In recent years I have had the pleasure and the challenge of talking with many students, reporters, and others with environmental concerns who often ask the question: *"Are Pesticides Really Necessary?"* They had read about genetic, biological, and other nonchemical methods of control and at times had been showered with reports of chemical hazards to health and the environment, some real and some highly problematical. Our discussions convinced me that, for the most part, these people were quite unaware of the proven benefits of pesticides; they had been shown only the negative side of the coin. *Wise decisions on trade-offs require knowledge of both sides of a controversy.* I hope this book will give the reader some insight into the benefits of pest control chemicals which must be balanced with risks when those do, indeed, exist.

First, it is important to recognize that farmers must defend their crops and livestock by one means or another against a wide range of pests if they are to provide mankind with a constant and adequate food supply. History is replete with records of famines caused by various pests: locusts, since early biblical times (Exodus 10); epidemics of wheat rust through many centuries; and the late-blight disease of potatoes, which caused the great Irish famine of the 1840s. We must also defend ourselves against organisms that destroy our structures and our stored food. If we are to avoid malaria and other diseases that have caused widespread misery, defenses against the transmitting insects must be maintained on a continuing basis. To treat the disease after the infected vector has bitten is "locking the barn door after the horse is stolen."

Fortunately, nature provides defenses against many pests. For instance, cabbage may be devoured by loopers, but most other garden plants are resistant to this insect. Plant breeders have gone a long way in intensifying natural resistance; they have developed rust-resistant wheat, for example. There are now some good biological controls and cultural practices that mitigate diseases, insects, weeds, and nematodes. Well-built structures protect stored products from mold and rats, while sanitation and other environmental control practices check populations of rodents and agricultural pests. But after these methods have all been put to work, there still remains a wide gap between the degree of protection they afford and the safeguards man needs to assure his food and fiber supply, to protect his health, and to conserve his resources. During this century pest control chemicals have provided that added measure of defense.

First came the inorganic chemicals, such as the arsenicals for insects and copper sulfate for plant diseases. There were also complex organic compounds extracted from certain plants that, like medicinal herbs, have special ability to synthesize biologically active materials; pyrethrins from the daisy-like pyrethrum plant, for example. Then it was learned how to synthesize other complex materials which chemists classify as organic (because, like the molecules that make up plants and animals, the major building block in their composition is the element carbon).

Today, there exists a wide range of chemical tools, often called pesticides, from which the best can be identified for a given task. Those commercially available have been chosen by manufacturers and regulatory officials for reasons of efficacy and safety when used according to label directions. Let's consider some important aspects of pesticides that should be brought into focus before trade-off decisions are made.

Part I of this book is an examination of the many defenses—natural and man-contrived—that keep pests from overwhelming us. Part II describes a number of situations where available biological, genetic, and cultural controls have not proven adequate and where pesticides have been needed for health protection and the avoidance of unacceptable losses in farming and forestry, as well as to prevent the destruction of stored products and wooden structures. Part III deals with the question of pesticide safety to people and the environment. The book will also consider the emotional, political, and scientific aspects of recent pesticide controversies.

In these three parts I shall stick to the facts as I know them or as others with appropriate qualifications have reported them. No doubt my personal views are implied here and there, if not specifically expressed, but I have seriously attempted to confine my own opinions to the pages following Part III under the heading "Trade-Offs."

The reader should know of my possible biases. I recently spent several years conducting research for a chemical company, my goal being the development of new and better herbicides. The several herbicides I discovered or encouraged my employers to develop and manufacture were thoroughly evaluated, prior to commercialization, not only for their human and environmental safety but also for their environmental benefits. I am proud of the fact that they have helped make farming and forestry a more productive endeavor and thus have reduced the amount of land that must be devoted to raising life's necessities. More land for recreational areas and wildlife preserves has thus been made possible. But we herbicide researchers are also proud of the positive environmental benefits of our products in terms of the reduction in energy required for farm and forest production, and in terms of reduced soil erosion.

Before I began doing research work with agricultural chemicals, I was for many years involved in plant breeding, especially in the development of rust-resistant wheats and in the breeding of asters for resistance to wilt—a soil-borne fungus disease. Later, while a faculty member at Auburn University, I engaged in research aimed at developing beans resistant to the rootknot nematode, and at Michigan State University I participated in a project which sought to combine borer resistance with other desirable characteristics in sweetcorn. Through this research I gained a keen appreciation of the important role of the plant breeder. Probably no tax dollars have paid off more handsomely than those that have funded the breeding of crops for disease resistance. Today, efforts continue—in the seed industry and at tax-supported research stations—to develop superior pest-resistant plant varieties for the benefit of man.

I may therefore lean favorably toward the use of chemicals, and I definitely support the application of genetic controls when they are available. I think I also have some bias in favor of biological controls wherever they are effective. I don't exactly enjoy spending money on pesticides or taking the time to apply them if there is an easier way. I refuse to grow anything in my garden that is too demanding.

My one overriding concern is that all countries have an efficient agriculture with high yields and, through application of the best available methods of control, with minimum losses from pests. This is another way of saying that I am in favor of abundant food production, year in and year out, and that I hope no one will have to go hungry and no shortages of any important food will stoke the fires of inflation. I favor high yields because they are essential to adequate nutrition for the world's growing population, and they minimize the need to expand farming to hilly land subject to erosion. Further, high yield crops tend to require a lower energy input per unit of production.

There is no way abundance can be assured, unless measures of one kind or another are taken to control ever-present pests. Man must either protect his crops and livestock or revert to the vagaries of food supply found in nature. Our forests, our structures, our clothing, and our stored products must be protected against insects—as an energy-and-natural-resource conservation measure, if not on immediate economic grounds. Finally, if we are to avert the horrors of malaria, encephalitis, and bubonic plague, we must control pests that transmit infections.

Yes, we must defend ourselves from pests if the kind of life that most of us want is to be realized. Can nonchemical methods alone provide adequate defense or are chemicals still essential to effective integrated pest management programs? In recent decades pesticides have contributed significantly to this defense, but like other tools, they carry some measure of risk. I hope the following pages will contribute to the understanding needed for wise decisions on trade-offs.

# PART

# I

## *Why Pests Have Not Overwhelmed Us*

[A look at the many natural and man-contrived
factors that keep pest populations in check]

# 1

## *Man and Nature Working Together*

Termites destroying your home . . . Worms devouring a farmer's cabbage . . . Mosquitoes transmitting encephalitis . . . Rust debilitating wheat and blight destroying potatoes . . . Internal parasites putting a cattleman out of business . . . Nematodes infesting the roots of tomatoes in your garden . . . Maggots in cereal products, in storage or on your kitchen shelf . . .

**Few of these calamitous things happen today because man has learned how to defend himself against the ravages of pests.**

Many writers have alarmed the public with alleged facts and sometimes fiction regarding the overwhelming menace of pests to man's food, his health, and his general well-being. Recently, *Reader's Digest* painted a gloomy picture in an article, "The Bugs Are Coming." This title, along with the magazine's first-page statement—"Humankind has made little progress in its age-old battle against insects"—bring forth a vision of the world being overwhelmed by billions of tiny King Kongs. Fortunately, subsequent pages were less gloomy than this exaggerated opening sentence. Nor can I be gloomy when I think back fifty years and recall how grossly inadequate our insect and disease controls were then, compared with today's vastly better, though still imperfect, pest management systems. Try to tell farmers whose experience spans several decades that we have made little progress in insect control. There is Charlie Wilson, a Michigan potato grower with yields upward of 300 sacks of a hundred pounds each per acre, who remembers when every year leafhoppers so debilitated his crop that one-third to one-half of the potential yield was lost. And there is Jack Marshall of East Texas, who no longer suffers livestock losses from screwworm infection of his cattle, because of a nationwide biological control program and good insecticides for emergencies. His cattle grub problem, too, has all but disappeared after several years' usage of one of the "pour-on" systemic insecticides.

Try to make this "little progress" story convincing to an old-timer in termite country who remembers the days before wood preservatives and insecticide-treated barrier strips around buildings came into being. Or try to tell me! I once had a promising head lettuce business "go down the drain" because of a plant virus transmitted by the six-spotted leafhopper for which we then had no practical control. One of today's approved insecticides would have kept me in business.

Actually, nature provides us with much of our defense. Widespread genetic resistance, weather adverse to the pests, predators, and parasites—all these contrive to keep pests from overwhelming us. Of course, we have been overwhelmed at times. Example: the rust of coffee in Ceylon during the nineteenth century. Incidentally, growers then turned to tea as an alternate crop, and Britain—which gave import preference to her colonies—became a country of tea drinkers by economic necessity.

Let's get back to some overwhelming pest outbreaks. The Irish Potato Famine of the 1840s caused by the late-blight fungus *Phytophera infestans* is estimated to have resulted in a million deaths and induced at least a million more people to migrate, largely to America. Less shattering to the lives of people and the course of history, but locally disastrous, was the outbreak of pea weevil, which ran the dry split-pea business completely out of my home state of Michigan in the early part of this century. Peas for your soup now come from western states where this insect is not a severe problem.

But coffee rust did not attack wheat, and the potato blight had no adverse effect on sugar beets. Corn and other cereals growing on the same farms as peas were unaffected by the weevil. When my lettuce was overwhelmed by the yellows virus transmitted by a leafhopper, the cauliflower in an adjoining field came through beautifully. Throughout history, man has experienced engulfing catastrophes from pests, yet he has survived thanks to the tendency of most of them to attack only a few hosts. By encouraging nature, by using certain cultural practices, and by supplementing these with the chemicals we call pesticides, the chances of our experiencing devastating losses have now been vastly reduced.

Among nature's many wonders, I marvel most at the specificity of pest-host relationships. If you are a vegetable gardener, you've no doubt had trouble with those pesky cabbage worms or loopers that attack members of the Brassica group of vegetables, but usually leave the rest of the garden untouched. Chinch bugs that raise havoc with your lawn stop their depredations at the beginning of the shrub and flower beds. The black-spot fungus on your roses disdains most other plants in your garden. There is some cross-species activity, of course, as you may have observed with cutworms or Japanese beetles, but specificity is more the rule than the exception.

If each kind of plant and animal did not have its own built-in defenses against most of the world's pests, we would all have long since perished (or more likely, life as we know it would never have evolved in the first place). Let's call these defenses genetic controls. Of course they have a biochemical basis. Sometimes a given pest is chemically attracted to a specific host.

Whether the pest is inadequately nourished by a less-favored species or cannot go through its life cycle in a given host environment or cannot gain entry to the host's tissues is all determined within the genes of both the host and the pest.

As the very fact of evolution teaches us, heredity is not a constant thing. Pests sometimes adapt themselves to new hosts through mutation and natural selection. On the other hand, genetic changes followed by selection sometimes give host species a new level of tolerance, at least until further hereditary changes in the pest enable it to again readily use the host as a source of nourishment and a home for one or more of its life stages.

Weather is one of nature's great protective devices. We may complain when it's excessively wet, but many potentially troublesome pest species are kept in check—directly by rain or because parasites are encouraged by humid conditions. Grasshoppers are a good example. When it turns dry, mosquitoes disappear but mites thrive. Most diseases of foliage are less troublesome when there is little rain or dew on the leaves in which spores can germinate, yet moisture stress seems to encourage some systemic fungus infections.

Yes, weather variables have a pronounced leveling effect on pest populations. Winter, with its cold soil temperatures and freezing and thawing, keeps nematodes in check as well as many insect species whose wintering-over stage is not highly tolerant of temperature extremes. There are many serious insect pests in the South that rarely reach troublesome populations in nothern farms and gardens. Bean beetle, corn earworm, and various worms infesting cucumbers, squash, and tomatoes are examples. Yes, there are advantages to farming and gardening in the North, even though by February we may have difficulty recalling just what those advantages are.

You've heard the verse about greater fleas having lesser fleas "and so on ad infinitum." And how fortunate that it's that way. Certainly life as we know it could not have evolved had it not been for parasites and predators. Man's many insect foes have their own enemies to contend with, including predacious insects such as ladybird beetles, lacewings, and others. Plant parasitic mites have to deal with these problems plus predacious mite species that feed on them. They live in a cannibalistic world. Nematodes are attacked by a range of parasites and predators; this, together with the freezing and drying of the soil, keeps these root-infesting parasites from overwhelming us.

Like man, insects are beset by infections from fungi, bacteria, and viruses. Bugs really have a tough time of it: additional problems are posed for them by birds and small reptiles and amphibians which spend their waking hours seeking out things that crawl or fly. Even mammals, including bats and other rodents, work to keep insect populations in check.

There is one more natural factor that plays a part in keeping pest populations from running wild, and that is food supply. Food for a pest species—usually a pest's specific host—is of course finite; when pest populations outrun their food supply the inevitable malnutrition, if not out-and-out starvation, results in a dramatic reduction in pest numbers. When we raise crops or livestock, a vast amount of food is created for specific pests, whose populations can explode unless other factors come into play; thus, great damage to agriculture can occur, but today man-contrived controls back up natural ones in keeping sources of human food from being overwhelmed.

Even in an undisturbed state of nature, populations of most species are seldom in the perfect balance sometimes depicted. I recall the vivid description of the so-called balance of nature given by my old high school biology teacher. He had a platform balance on his desk, one tray of which he labeled "factors favoring population growth" and the other "factors unfavorable to an increase in population." By placing adequate food, favorable weather, and living space on one side of the balance cards, he illustrated how a species would increase rapidly when these conditions prevailed, but then, as cards labeled parasites, adverse weather, and a shortage of food were placed on the other side, he showed how populations would be reduced. We could shift the delicate balance merely by adding or removing a card.

Man's many activities, including the agriculture he developed as a source of food and fiber, have often caused the balance to swing wildly, but even in a state of pristine nature, it is far from stationary and often moves through a wide arc. The very facts of evolution and ecological succession is evidence that imbalances often occur.

Living space itself can sometimes be a factor in pest populations. Aside from the effect of crowding on food supply and the spread of parasites, cannibalism and self-destruction often follows. We have all read of the result of overcrowding, of the suicidal migrations of lemmings in Scandinavia, for example.

The next several chapters of Part I are an examination of the many defenses man has contrived to supplement nature's way of keeping pests from overwhelming us: genetic, biological, mechanical, and cultural as well as chemical controls. Then, in Chapter 8 ("Putting It All Together"), integrated pest management systems using all these defenses—scientifically designed systems that are increasingly being employed to keep pest populations below damaging levels—are considered.

# 2

## *From Genes to Rust to Hessian Fly*

Some of nature's techniques of keeping pest populations from getting out of bounds can be encouraged and intensified by man. Let's look first at our successes in reducing the severity of pest damage by breeding for resistance. Since the time agriculturists first purposefully saved seed from the best plants or identified the healthiest livestock for breeding purposes, there has been considerable man-directed selection for resistance to diseases. This was essentially natural selection speeded up by man. He had only the variability that nature gave him to work with, but by perpetuating the healthiest individuals, resistance was often intensified at a much greater rate than if left to natural selection alone.

While early agriculturists had to be content with natural variability, over the past century man has devised a number of schemes for increasing diversity and for combining desirable traits from different strains in one variety of breed. Genetic manipulation has become a highly scientific endeavor.

Wheat, the crop responsible for our daily bread for the last six thousand years, is attacked by the rust fungi, so called because of the rust-colored pustules they cause on leaves and stems. This group of plant pathogens was responsible for many of the famines of history, and were it not for the scientific combining, in one variety, of resistance to various races

7

of the rust organism, carried out by plant geneticists of this century, wheat yields would still be drastically reduced in the years favorable to the fungus.

Grains from badly infected fields are often shrunken and of poor quality for making flour. The story of wheat rust and other fungus diseases throughout history, depicted in the book *Famine on the Wind*,* makes interesting reading.

Modern plant breeders are now combining resistance to many different diseases in a single variety or hybrid. Thumbing through a seed catalog, I note the letters VFN after several tomato varieties; according to a footnote, this means that a particular variety or hybrid is tolerant of verticillium and fusarium wilts and also nematodes. Tomatoes are notoriously susceptible to a number of diseases, but many of today's fresh market and processing varieties have resistance to more than one disease. There are few crops that have not benefited in recent decades from scientific breeding for pest resistance.

Resistance is seldom absolute; breeders rarely achieve complete immunity. Other methods of control are sometimes needed to supplement tolerance to pests, but together they can result in a higher level of productivity than through one method of control alone. Years ago, while on the Agricultural College faculty at Auburn University, I participated in research on a nematode-resistant variety of bean. Grown in soil heavily infested with the rootknot nematode, it would produce a reasonably good crop where nonresistant varieties would fail almost completely. Yet, we found that, on fumigating this soil with chloropicrin for nematode control, our resistant variety would produce almost twice the yield that it did where the nematodes were not killed in advance of planting.

Much depends on the infection pressure prevailing at a given time. Many resistant or tolerant varieties produce well

*G. L. Carefoot and E. R. Sprout, *Famine on the Wind: Man's Battle against Diseases* (Rand McNally and Company, 1967).

when the pest population is modest or, as in the case of leaf diseases, when conditions for spread of a fungus or bacterial infection are not highly favorable. Even though resistant varieties may outperform susceptible ones, they may still suffer considerably if infection pressure is very high.

Man's greatest accomplishments in modifying genetic resistance have been with the fungus diseases, while more limited success has been achieved with those diseases caused by viruses and bacteria. The more difficult job of breeding for resistance to insects is only in its infancy, although some remarkable progress has already been made with several farm crops—for example, wheat resistant to Hessian fly, potatoes to flea beetle, and onions to thrips.

If you were keeping up with the news in August, 1970, you may recall the great concern in the United States regarding the unprecedented spread of corn blight, a fungus disease ordinarily confined to the South which suddenly moved northward on the wind. Encouraged by unusually wet weather over a large area, the disease lowered corn production that year by 15 to 20 percent or over a billion bushels. In spite of this reduction in yield, the 1970 yield of 71 bushels per acre was far ahead of the yields in the decades prior to the development of modern corn technology (*see* Chapter 24).

Corn breeders and pathologists soon determined that a particular gene carried by much of the corn crop and useful in producing hybrid seed was also associated with susceptibility to the southern blight disease. Crop specialists began asking themselves, "If severe blight could occur on corn because much of the crop carries a gene resulting in susceptibility to a specific disease, how about other crops? Have we unwittingly got ourselves into a vulnerable position by having too much of the same genetic composition in our major food plants?"

Much has been written about genetic vulnerability, a term frequently used following the 1970 corn blight, and there has been much criticism of agriculture for its having employed too narrow a range of germ plasm. The 1970 corn blight epidemic may have been a blessing in disguise. We certainly do not want our major food crops to be excessively vulnerable to disease, and now more attention than ever is being given to this problem by plant pathologists and breeders everywhere.

In reality, economic plants have always been genetically vulnerable to disease, even when man still depended on gathering seeds and roots from the wild. Genetic diversity *per se* does not provide resistance or even give assurance that a significant part of the total species population will survive an epidemic. Look what happened to the American chestnut in the early part of the century and what is happening to elms today! There have been widespread outbreaks of plant diseases throughout history; from biblical times—when food shortage in Egypt may well have been due to wheat rust—to the Middle Ages when that disease definitely is known to have caused many famines in Europe.

There was plenty of genetic diversity in times past; then, farmers tended to save their own seed, and there was no seed industry to distribute improved varieties. Neither were there privately and publicly-supported plant breeders collecting different germ plasm far and wide, cataloging it for resistance factors, and then combining a multitude of desirable traits in one variety through hybridization and selection.

Today, there are many "germ plasm banks" where a tremendous number of different varieties of various crops are maintained specifically for use by plant breeders now and in future years. When I was a student at the University of Minnesota in the early 1930s, a large collection of rust-resistant wheats was being maintained by collaborating agronomists and plant pathologists. One of my part-time jobs was helping raise small plots of these varieties, often from the far corners of the earth, and then storing seed samples under ideal conditions for longevity.

Recently I had the interesting experience of visiting the

outstanding seed bank maintained by the U.S. Department of Agriculture at Fort Collins, Colorado, and on another occasion the privilege of viewing the world's largest collection of manioc or cassava at a research station in Columbia. Stored in refrigerated concrete chambers at the International Maize and Wheat Improvement Center near Mexico City are more than 13,000 samples of corn (maize) seed collected from around the world, and ready for geneticists to use in their continuing efforts to breed commercial varieties and hybrids with resistance to disease. The International Rice Research Institute in the Philippines, in an effort to conserve the world's traditional rices for use by future plant breeders, has collaborated with a number of countries in collecting varying types from farmers' fields. The germ plasm bank at that institution now contains more than 50,000 accessions. (Viable seed is replenished by growing small plots when the stored seed begins to lose its ability to germinate.)

In all, I know of collections and seed banks for over 30 of the world's important crops, and there are undoubtedly others I have not heard of. Since the mid-1970s, a branch of the United Nations Food and Agriculture Organization in Rome has coordinated and supported efforts to collect, preserve, and make available to plant breeders everywhere the germ plasm of important plant species.

Do you wonder that we old-timers tend to smile when we read criticisms of modern agriculture implying that little attention has been paid to genetic diversity? Yes, farmers raise a lot of the same variety because it has proven productive and fits their needs. But for major crops the breeder is there, with his arsenal of diverse genetic material, ready to utilize different genes for resistance as the need arises. Seed banks and crop diversity will be discussed further in Chapter 32.

# 3

## Making Nature Work Harder

When man began to move more freely around the globe, he inadvertently carried with him a wide range of pests, but more often than not, the natural predators and parasites that kept the pests in check on their home ground failed to keep pace. It seems almost axiomatic that, when a pest is first transported to some far-off part of the world having an environment favorable to its reproduction and survival, there is a population explosion because natural enemies have not been transported with it. Rabbits from Europe practically ate up the countryside of Australia and New Zealand until this mammal's natural diseases were imported by man and became well established. The fungus causing the Dutch elm disease is a current example of an imported pest with which we have become all too familiar in the United States in recent years. Unfortunately, attempts at checking the elm bark beetle that transmits the disease have shown only modest success. When I was a boy, the European corn borer caused terrific damage, but now, since its natural enemies have also been imported and become established, the corn borer is just another insect, one that must sometimes be controlled by insecticides, but often is not present in damaging numbers.

During the latter decades of the nineteenth century, the fruit tree insect known as the cottony cushion scale was unwittingly introduced into California. This pest, which attaches itself to the bark of every twig and extracts plant sap for its nourishment, killed hundreds of thousands of citrus trees and threatened the entire orange industry of that state. Entomologists studying the problem learned that the scale originated in Australia, where it seldom caused significant tree injury, apparently because its natural enemies keep numbers in check.

So the United States government sent an entomologist to Australia to search out the parasites and predators that might be considered for importation into California. He chose a species of ladybird beetle because, in itself, it was unlikely to become a pest and was not believed to feed extensively on other desirable insect species. The 500 beetles imported multiplied prolifically and—within two years—brought under control the devastating cottony cushion scale, which since then has seldom again been a serious pest in California. The Australian ladybird beetle has subsequently been shipped to many other countries, where it has generally been successful in controlling that same imported scale insect.

There are a number of other success stories of how outbreaks of exotic species have been brought under control, including accounts of how several parasites of the European corn borer, the gypsy moth, and certain mealy bugs, weevils, and aphids were established. But these control methods are seldom perfect; sometimes, populations of the damaging species get way ahead of the parasites or predators and insecticides are required to cut back the numbers to manageable levels. More often, the general population level of a given insect is kept down by the predator or parasite, as in the case of the bacillus causing the milky disease of the Japanese Beetle and other beetles, the larval stages of which can damage our lawns. As destructive as the Japanese Beetle can still be in some instances, it was far worse in the 1930s before its parasites had been identified, propagated, and disseminated. Natural enemies have been found useful against weeds in a few instances; for example, the Klamath weed in the western United States has been controlled by a leaf beetle imported from Australia.

The biological control most widely used by farmers and gardeners is *Bacillus thuringiensis,* an insect pathogen effective against cabbage loopers and other larvae of the moth and butterfly group of insects known to entomologists as the Lepidoptera. Spores of this bacterium are now formulated under various trade names and, after dilution according to label directions, are sprayed like an ordinary insecticide. Action is not as fast as with some chemical insecticides, but under many conditions, this biological agent has proven effective, particularly if applied when worms are still quite small.

Considerable progress has been made in recent years in identifying and multiplying other pathogenic bacteria and also viruses that attack insects. The extensive research still being undertaken in this area will likely result in further control of various pests by the dissemination of disease organisms. If products containing these biological agents prove safe and gain approval by governmental agencies, they will no doubt be prepared in commercial quantities and find significant places in pest management programs.

Considerable efforts have been made to culture predators such as certain species of ladybird beetles and parasitic wasps for release in numbers adequate to control particular pests, such as aphids, weevils, beetles, and scale insects. Today, this technique has become of considerable interest to gardeners and fruit growers, and has often been successful. The staggering number of predators that must be cultured and released for effective insect control in large-scale farming operations have seldom been available, but given the right conditions, some species will multiply rapidly enough to contribute significantly to control.

Much research is underway at the present time on improved methods of culturing desirable insects for large-scale release, and further progress on this approach to pest control will undoubtedly be made.

No conniver bent on playing dirty tricks on his adversary has ever equalled the "treachery" of the entomologists in their male-sterile approach to insect population control. In seriousness, this manipulation of nature has had some successes and promises more for the future. It all started a few decades ago when new knowledge of the adverse effect of radiation on the fertility of male insects and the similar loss of fertility following exposure to certain chemicals became available. Some imaginative entomologists speculated that, if massive numbers of male insects could be sterilized and then released in areas where mating normally occurs, they would dominate the scene and greatly reduce the insect birth rate. The chances that this travesty of nature would succeed were most favorable among insect species that mated but once.

One of the early success stories with this scheme involved a massive attempt to eradicate the screwworm from the southern United States. The larvae of this fly infest the flesh of all kinds of livestock after the adult females deposit their eggs in scratches or wounds. After millions of sterilized males were released over large areas, the screwworm was seemingly eradicated, but unexpected outbreaks later occurred, causing livestock producers to check their cattle carefully and to treat the cattle's wounds with an appropriate insecticide for the control of the larvae. These outbreaks of screwworm—after eradication was thought to have been achieved—and the necessity of using available livestock insecticides to avoid severe losses emphasize the contention of many that we should always have more than one pest control measure ready to go into action. Currently, efforts are being made to control the cotton boll weevil by the sterile-male-release technique. No doubt this approach will make useful contributions to the management of other destructive species in future decades.

Biologists continue to study the life cycles and physiology of pests as well as parasites and predators, hoping someday to translate this basic information into new approaches to control. A recent development is the inoculation of plants with attenuated strains of a virus or fungus to compete with more virulent forms and thus prevent the forms from overwhelming the host. Distinct species of pathogens which have little adverse effect on the host are proving promising for minimizing the effect of more virulent forms. (Some researchers think these less harmful fungi synthesize antibiotic substances, which then inhibit the more serious pathogens from gaining an upper hand.) Also promising is the encouragement of certain nonpathogenic soil microorganisms that tend to be antagonistic to the development of parasitic species. And in cases where some species of plant parasitic mites were found to have developed resistance to insecticides, a similar resistance was brought about in beneficial predaceous mites to help them control the destructive strains; this involved the deliberate selection (and subsequent release) of predatory mites for tolerance to one or more insecticides essential for a specific crop in a given area. Possibly other genetic improvements in beneficial mites and insects will give us one more useful defense to integrate with other methods of control.

# 4

## Some Tricks of the Trade

We have unconsciously solved many of our pest problems simply by not attempting to raise specific plants in areas where they are beset with insects or diseases. The shift of dry peas from Michigan to the West has already been mentioned as an example. I make no attempt to grow potatoes in my garden because of the inevitable problems with beetles, leafhoppers, and leaf blights; it's more practical for a potato farmer with a sizable acreage to control them. Large acreages of winter and spring onions are produced in Texas, but not in Florida, partly because of the severity of downy mildew under the damp conditions in the Southeast and the usual lack of this disease under the drier atmospheric conditions of Texas. These and other mechanical and cultural controls, Tricks of the Trade, go a long way toward minimizing pest damage.

Disease-free seed is an important factor in control of viruses and bacterial diseases; seed is therefore often produced in areas where their incidence is relatively low. For example, seed potatoes are grown in northern areas where viruses are less prevalent, and bean seed is grown in the western United States, where anthracnose and bacterial blight are seldom present. The new tissue-culture technique of propagating plants has made it possible for nurserymen to obtain disease-free stock of a range of vegetatively propagated species; chrysanthemums, raspberries, and strawberries are examples. To maintain these stocks in a disease-free condition, nurserymen sometimes grow plants to be used for further conventional propagation in screened houses free of virus-transmitting insects.

When you plant can be just as important to the avoidance of certain pests as where you grow the crop. Long ago it was found that, if winter wheat was seeded too early, larvae of the Hessian fly could do great damage to the young plants. In the North, seedings made in October avoid this problem. Spring-planted radishes, cabbage, and other members of the Brassica group of vegetables are often troubled with root maggots, the larvae of a fly that lays its eggs at the base of the plant. But this is largely a spring problem. If you want these vegetables early, you will have to use some other method of control, but for a relatively maggot-free crop just plant in June or later.

I recall the admonition of a farmer for whom I occasionally worked when I was a boy: "Never cultivate beans early in the morning." I doubt that he knew why, but there was more than superstition to this folklore. A bacterial blight of beans which causes tiny lesions on the leaf can be readily spread when the plants are wet with dew. The disease often shows up first in only one spot in the field, but a cultivator—in those days involving a horse pulling it and a man following on foot—can readily spread the tiny bacteria around the field.

I also recall the practice of my grandfather to plow down all of his old cornstalks by early May. This was hardly folklore; it was recommended, in fact sometimes required, by government agents who were attempting to prevent the spread of the European corn borer. The pupae or resting stage of this insect resides in stalks all winter. If the remains of last year's crop are buried early enough in the spring, the adult moth cannot emerge, later, to deposit eggs on young leaves which promptly hatch into the destructive worm. Plowing down cornstalks is not as essential these days, because natural enemies of the insect tend to keep it in check and the populations are usually not as great.

Sanitation is an important aspect of pest control. Rodents thrive and will increase in numbers where food is present. No other control measures can be meaningful if garbage and other sources of food are available. Prompt spreading of barnyard manure during the fly season has long been known to be important in the control of those fly species that go through a part of their life cycle in animal dung. The wise fruit grower who sees an initial outbreak of fire blight of apples or pears immediately prunes off the affected limbs and burns them, to minimize the spread of this bacterial disease. You would be an unwelcome guest at a tomato greenhouse if you came in smoking. The tobacco mosaic virus also affects tomatoes, and it can be transmitted by smokers. The elimination of an intermediate host can sometimes reduce the incidence of a fungus disease. Example: a species of currant which harbors one life stage of the white pine blister rust.

Mechanical exclusion of pests or the modification of environments so that pests will be less likely to prevail has always been an important factor in pest control, particularly in the protection of stored products. Rodents must be kept out of storage areas by proper storage construction, and rain must be excluded. Damp grain or other stored products will support the growth of fungi and bacteria, and quickly spoil. You may have noticed a sheet of aluminum or galvanized steel laid down between the top course of blocks in your house foundation and the wooden sill or plate. This barrier is part of a termite control program. It will not do the job by itself, but it does make it more difficult for termites to build their mud tunnels from the soil up to the wooden parts of your house. Gardeners often place a screenwire collar around young trees and shrubs to prevent mouse and rabbit damage to the bark.

Mechanical methods of actually killing pests have long been important, and the fly swatter and sticky flypaper are well-known extermination tools. The flypaper idea is also proving useful for monitoring purposes; for example, round wooden spheres painted red to simulate apples and coated with a sticky substance will catch adult apple maggot flies when hung in trees. The population counts this makes possible tell the grower when spraying with an insecticide will give the most effective and profitable maggot control. A commercial sticky substance spread in a ring around the bark of your shade trees prevents the wingless females of canker worms from crawling up the trunk and reaching the leaves where they will eventually deposit their worm-producing eggs.

Another "sticky-trap" idea is being used for keeping populations of greenhouse whiteflies from reaching damaging numbers. This pest is attracted to a deep yellow surface, and—according to a recent report from the U.S. Department of Agriculture—pieces of boards appropriately painted and then coated with a heavy oil or commercial insect trapping compound will attract them when placed between infested plants. The method is useful for preventing build-up of numbers when only a modest population exists, or following the reduction of severe infestations by an appropriate insecticide.

Other trapping schemes have been used for various insects; for example, lights to attract certain photosensitive species to their chemical or electrical demise. Light traps have often proven useful for population monitoring, even where they are impractical for use over large areas. (The use of chemical attractants, including pheromones, will be discussed in Chapter 7.) Trapping has long been important in the control of mice, rats, and other vertebrate pests.

Environmental control or modification is often an important factor in the prevention of serious pest infestations. These days, poultry houses have insulated ceilings to reduce the drip of water condensate during cold weather and thus to help prevent respiratory diseases. Animal shelters of all kinds are, of course, an attempt to control the environment. An unusual environmental modification, the occasional emission of high frequency sound, shows promise in some experiments as a method of driving certain rodents and insects away from adjacent premises. Prolonged flooding, a truly drastic change in the environment, has been used to reduce populations of certain soil-inhabiting fungi and other pests.

Greenhouse operators raising tomatoes keep their boilers going at a slow pace, even in late spring, because growers have learned that if they heat just enough to prevent dew formation on the leaves, the incidence of leafmold, a fungus disease, is greatly reduced. Environmental control of mosquitoes through drainage of stagnant water and the elimination of cans, bottles, old tires, and other objects that collect rainwater is an important step in the effective management of this pest.

Various modifications of cultural practices—for example, the proper choice of timing and method of irrigation to avoid conditions favorable for the spread of plant diseases—can aid in minimizing pest damage. Adjusting plant spacing sometimes lowers the incidence of fungus infection: foresters have learned that Hypoxylon canker of aspen, now a major pulpwood species, can be reduced by maintaining dense stands.

The agent at a port of entry who asks whether you have any fresh fruits, vegetables or live plants in your luggage isn't trying to be perverse. Along with those who inspect commercial shipments for unwanted insects he is an important cog in man's effort, through quarantines of one kind or another, to prevent further spread of pests. At this writing, the citrus black fly has jumped the international border control and has gained entry into South Florida. Now a new quarantine line has been set up in an attempt to confine the pest and hopefully to eradicate it from the United States. If the black fly should spread northward to the major Florida citrus belt, production could be drastically affected, and your breakfast grapefruit or orange juice would inevitably become more expensive. Entomologists of the University of Florida are presently making good progress in identifying and culturing insect predators of the citrus blackfly, and there is optimism that effective biological controls will eventually become available. Meanwhile, preventing or reducing the rate of spread can prove of great value to both producers and consumers. Recently the Mediterranean fruit fly was discovered in the Santa Clara Valley of California and great effort is being made to eradicate this very serious pest.

# 5

## Crop Diversity – Some Contributions and Limitations

Crop diversity and rotations are sometimes helpful in reducing the severity of pests. Many soil fungi and cyst-forming nematodes, however, remain active over a long period, and rotations are not always practical. For example, the sugar beet nematode can be controlled by long rotations in growing areas in the eastern United States, because a relatively small part of the land is farmed to this crop, and growers can readily shift their beets from one field to another. In the inter-mountain valleys of the West, sugar beets are a major crop, and there is not enough land available to permit the long rotation required for control of the beet nematode; thus, soil fumigation is often required.

Some soil insects, e.g., the corn rootworm, are reduced in population by rotations. However, there may still be enough insects present to require the use of additional methods of control for top yields. Rotations alternating row crops with a sod crop consisting of grass and a legume used for hay or pasture have advantages from the standpoint of biological nitrogen fixation and maintenance of good soil tilth, but they are not always a panacea for soil-borne pests; some insects, such as cutworms, tend to be present in greater numbers following the sod phase of a rotation. Maintaining a high level of organic matter through the use of cover crops and the application of manure does help keep some kinds of nematodes in check, apparently by supplying energy to organisms that parasitize them.

The notion seems to prevail among some environmentalists that, with adequate diversification, one can achieve practical pest control just by letting nature take its course. In an article in *Environment* for April, 1970, Barry Commoner referred to the desirability of ". . . a new, more mixed form of agriculture that will make it possible to get rid of most insecticides and make better use of the natural biological controls." A more diverse agriculture would minimize some insect pests, but others would in all likelihood be little affected. There is little evidence that the need for man-managed control systems would be reduced by any practical level of diversification, and these systems would often require pesticides as one important phase.

With biological as well as chemical control programs, successful pest management depends on diligent attention to detail. Sterile male insect releases, the employment of bacterial diseases of insects, and the culturing of predatory species all require a high level of technology, just as do the safe and effective uses of insecticides. Simply planting an array of different crops and then letting nature take its course would be a sure road to food shortages.

We have a summer place in western Michigan with enough room for a small garden and a lone apple tree. The surrounding land is wild and includes a wood lot and brushy areas. There are no other gardens within a quarter of a mile. Some years the apples—which I never spray—are clean, but in others they are riddled with maggots. I've tried growing cabbage and broccoli without an insecticide, but the worms usually win, even though there are no commercial plantings of these vegetables in the area.

Farming in the region—with deciduous fruit, canning beans, asparagus, field corn, and alfalfa—is highly diversified. These crops are well interspersed with unused land, woodlots, and pastures. In spite of this remarkable diversification, I have been unable to see, during the thirty-five years I have been spending holidays and weekends in the area, that local growers have fewer pest problems than in regions devoted to only a few kinds of crops. They must be alert to outbreaks of asparagus beetle, and codling moth would soon put the apple producers out of business if they did not use controls. Fireblight of pears and cherry leaf spot are serious when weather conditions are right for these diseases. The area is about as far from being a monoculture as any food-producing region I know of, but without diligent pest control the local farmers could make no more than a pitiful contribution to our food supply.

Trap crops (plants that specific insects like best), have sometimes been used successfully to attract destructive species, which can then be destroyed, mechanically, or by the use of an appropriate insecticide. An example is lygus bug, a particularly destructive pest of cotton in central California. This insect prefers alfalfa, and recent experiments indicate that strips of this hay crop in fields of cotton or other lygus-susceptible species may be used for cornering the bulk of the population. If a whole field of alfalfa is cut during the lygus season a myriad of adult insects may move over to nearby fields of cotton or seed crops, where they do more damage than to their preferred host, because they attack flowering parts of these other plants and thereby greatly reduce yields. One method of minimizing their migration is to cut alfalfa fields in strips at different periods so there is always some of this preferred plant to migrate to.

I have had success with my own trap crop scheme for our garden: a row of very-early-planted radishes or turnips to attract the fly-like adult cabbage root maggot, which lays its eggs at the base of members of the Brassica group of vegetables, including cabbage, cauliflower, and broccoli, in addition to radishes and turnips. When plants of the former three are set out after one of the Brassica root crops are well established, they are often less troubled with this spring pest. Of course the radish or turnip roots may be riddled by the maggots that develop from the eggs laid by the adult fly, but the space can be useful for tomatoes or some other early summer planting. As mentioned previously, late planting is another way to minimize maggot-damage to roots of these vegetables. (Remember that radishes and turnips can serve as traps only for maggots, and do not reduce damage to the Brassica vegetables from various leaf-eating worms. I have found no practical way to control these worms except by *Bacillus thuringiensis* or an appropriate chemical insecticide.)

# 6

## Some Facts and Fancies

Several old wives' tales proclaiming remedies for human ailments have stood the test of scientific investigation and led to useful drugs. How about those relating to the control of garden pests? A few of these old wives' tales have been found to have a sound foundation, but many remain in the realm of superstition. Like medicinal folklore, some have proven to be effective, many are yet to be scientifically investigated, and others are pure myth. Let's look at the marigold story.

According to legend, plants growing where marigolds grew the year before or where marigolds are interplanted often exhibit superior vigor. Workers at the Connecticut Agricultural Experiment Station have found that a root exudate from marigolds inhibits the activity of the root-infesting meadow nematode, so where this pest is present the benefit of marigolds can actually be realized. Dutch scientists who became interested in this subject have isolated and identified the specific organic chemical that the marigolds synthesize in their roots. This compound has not been commercialized as a nematicide because others of proven safety, when properly used, are more effective and more readily manufactured. Contrary to the impression often left by garden writers, the exudate of marigold roots is not effective against all nematodes; those species causing root-knot show little response.

Now, enter the imaginative garden writers. Control of all

kinds of insect pests seems to be attributed to marigolds. For the most part, these claims have not been subjected to scientific scrutiny. My own "clinical" observations indicate they are mostly myth. When proven otherwise, I will apologize to anyone who presents positive results from scientifically conducted experiments. Meanwhile, my roses will have to receive an appropriate pesticide when aphids or chaffer or mites get out of hand. Interplanted marigolds didn't do the job. The year there were claims in the garden press that this flower repelled rabbits was the same year this cute but sometimes destructive form of wildlife ate my French marigolds down to the groundline.

Over-imaginative garden writers have attributed all kinds of insect control benefits to companion plantings. A feature story in the gardening section of a southern newspaper stated, "Radishes encircling zucchini will help repel insects, and garlic will protect everything, from roses to rutabagas, from hungry pests." I have found no reports of experiments that show radishes or garlic will prevent insect damage to adjacent plants.

This mythology of repellent plants is not confined to reporters serving the popular press. Even some publications one would presume to be scientific have succumbed. In *The Unfinished Agenda*,* a report of a task force of environmentalists, one finds the statement that one way some farmers and millions of backyard gardeners control pests is by "...interspersing plants with insect-repellent properties such as marigold, garlic, onion and herbs." Correspondence directed to the editor of that publication and members of the task force, mostly university professors, who were responsible for writing the chapter in which this statement was made failed to bring forth reference to a single scientifically conducted experiment on the repellency of interplanted species.

The National Wildlife Federation has joined the repellent

---

*Gerald O. Barney, ed., *The Unfinished Agenda* (New York: Thomas Y. Crowell Company, 1977).

plant mythologists in a booklet reprinted from the magazine *Ranger Rick*, in which a list of fourteen specific plant species were recommended for interplanting.* Examples: "Mexican bean beetle... Repel them with summer savory." "Tomato worms. Plant asparagus, marigolds or borage near tomatoes to repel them." Correspondence revealed that no references to experiments confirming efficacy could be cited by either the National Wildlife Federation or the Rachel Carson Trust for the Living Environment which was listed in the booklet as a source of information.

I did learn through correspondence with research entomologists about work at the University of Illinois in which marigolds and also several herbs alleged in the garden literature to have repellent properties were interplanted with cabbage, egg plant, and beans. The results were all negative; the interplanted species failed to reduce the population of several common insect pests of these vegetables as compared with the control plantings.

I am sure there must be something to the old wives' tales about garden insect control with repellent plants. No doubt specific insects are sometimes discouraged by certain plants, and I hope additional scientific research on this subject will be carried out. But these broad generalizations, particularly when made in publications by university people or a prestigious organization, can be misleading. I wonder how many of the writers who make such claims have actually grown the interplantings they write about and have had effective control of the pest they mention and, in a parallel planting without "antibug" plants, an infestation of the same insect species.

Among the garden facts and fancies dificut to sort out is the notion held by many that plants are resistant to pests if they are well nourished, particularly if the nutrients essential

---

*Laurence Pringle, *Fighting Pests with Pests*, Publication of the National Wildlife Federation (Washington, D.C., 1978).

for plant growth are derived from organic sources. There are probably some instances in which a vigorous plant can keep ahead of certain pests, but generalizations are certainly not justified. Potato late blight will hit a vigorous field just as readily as one that is poorly nourished. I have seen no evidence that the common insects attacking garden plants such as cabbage, roses, or beans will be any less prevalent where the crop is in a vigorous state of growth, regardless of the source of nutrients.

Organic matter is a very important component of soil, and recycling the remains of garden plants, leaves, and lawn clippings, directly or as compost, is a highly desirable practice. Certain kinds of nematodes, including the one causing rootknot, have been found to be reduced in population by the liberal application of organic supplements. Apparently, decomposing organic matter furnishes energy for predaceous nematodes and fungi that parasitize the harmful nematode species. Some soil-borne diseases are partially inhibited by high soil organic matter, while others are little affected. Regarding composts and manure as an alleged aid in minimizing foliar pests, in contrast to deriving the same basic plant nutrients from fertilizer, I have been unable to improve on a statement by the late Dr. John Carew, who for many years was Chairman of the Department of Horticulture at Michigan State University. I will quote him as follows:

> All organic matter must be first decomposed to a chemical state before it moves into the plant. Research has repeatedly shown that plant nutrients derived from compost or other organic material are no better nor worse than those from chemical fertilizers.
>
> Plants grown organically are supposed to be more resistant to diseases and insects. According to this theory, "well fed" plants are less likely to be attacked and are better able to ward off diseases. This claim generally arises from the observation that home gardens have fewer disease and in-

sect problems than commercial farms. This is true, but more because of the intensive nature of commercial farming and the market demands for blemish-free fruits and vegetables. It is not really due to the presence or absence of chemical fertilizers.

In many respects plants are like people; a well-balanced diet will prevent most nutritional disorders but have little influence on susceptibility to certain pathogenic diseases—such as chicken pox or influenza in man, or fusarium wilt, aster yellows or seed maggot infestations in plants. Whether the plant gets its nutrients from organic or chemical sources seems to have no effect on disease or insect attack.

There is no doubt that plants often grow better in soil well supplied with organic matter, providing there is not a shortage of one or more essential plant nutrients. Among the benefits of organic matter are: support of growth of desirable micro-organisms including mycorrhiza, gradual release of nutrients, improved rainfall infiltration and moisture retention, good root aeration, and better availability of minor elements. Decomposing organic matter also supplements the atmosphere's normal supply of carbon dioxide, a building block for photosynthesis, and thus at times may enhance plant growth, but few pests will be less prevalent because of its presence.

It has often been observed that some kinds of crops are more vigorous following or growing in association with certain species. Recent research indicates there is a specific chemical compound, tricontanol, derived from decomposing alfalfa, which stimulates the growth of certain plants. Further investigations may reveal the identity of other chemical growth promoters associated with decomposing organic matter. It would be surprising indeed if there were not plants other than alfalfa that release specific growth regulators into the soil.

Conversely it is well recognized that exudates, leachates, and decomposition products of some plants are detrimental to the growth of others, a phenomenon known as allelopathy. An example familiar to many gardeners is the poor plant development experienced near black walnut trees or on soil where this species of tree has been removed. Chemists have now identified a specific compound, juglone, as being responsible.

Leachates from certain plant litter, for example sorghum, have been found to inhibit growth of some kinds of annual weeds, and researchers are currently exploring ways to utilize this phenomenon as one phase of integrated weed control programs. Studies also show that certain weeds are allelopathic and can harm certain crops over and above their competitive effect. In such instances, intensive effort to minimize their numbers is required if crops are to produce to their full capacity. The identification of specific phytotoxic compounds derived from allelopathic plants may provide research chemists with new leads useful in their search for improved herbicides.

Specific chemicals produced by living plants—derived from their litter or produced on their decomposition—that inhibit pests other than weeds probably exist, and further research may identify some that could eventually prove useful in pest control. To date, the nematicide produced by marigold roots is the best known and one that can have practical value where the meadow nematode has infested a garden.

# 7

## *Mesquite, Mites, and Molecules*

The practice of applying various substances to plants or soil and to livestock—and even people—for the control of a number of pests goes back a long way. It is only during the past century, however, that man has had the benefit of relatively pure chemicals to supplement other forms of control, and only during recent decades that a wide range of effective compounds have been readily available to aid in minimizing damage from the many kinds of pests. Fine dust or ashes sprinkled on leaves of crops to discourage leaf-eating worms were among the earliest pesticides. Salt was occasionally used to suppress weeds in certain salt-tolerant crops. The insecticidal properties of dried flowers of the daisy-like Pyrethrum plant ground to a fine powder were discovered long ago, and we have since identified and learned how to extract the specific and complicated chemical, pyrethrins, which gave the flowers their insect-killing power. Rotenone, another complex insecticidal compound derived from the roots of the derris plant, was applied as a dust long before the science of chemistry permitted us to identify the active chemical ingredient of this natural product.

Copper sulfate and other preparations of sulfur as fungicides and arsenicals as insecticides came into use during the latter part of the last century. According to some agricultural historians, the value of copper sulfate (blue vitrol) as a fungicide was discovered accidentally in France when a grape grower, in order to discourage theft by passersby, sprayed a solution of this blue compound on his vines growing near the roadside. The severity of mildew, so destructive to European grapes in many years, was greatly reduced.

These inorganic compounds were far better than nothing, but high dosages were required, compared with their synthetic organic successors, and were not always very effective. I recall the limited number of chemical tools we had for a wide range of crops when I was a boy; the arsenates for worms and other chewing insects, pyrethrum dust or a nicotine extract of tobacco for aphids and other sucking types. A neighboring greenhouse operator also used derris dust, a crude form of rotenone, for certain pests, and a fruit grower nearby used—in addition to these materials—an emulsifiable oil for a dormant treatment to control scale insects on his trees.

The new era of a wide range of specific compounds for specific pests was just dawning a half-century ago. Paradichlorobenzene had been used for clothes moths and someone found that this vaporizing synthetic compound was effective for peachtree borers. Then, in the 1930s, with the discovery that some substituted phenolic compounds had valuable insecticidal as well as fungicidal and herbicidal properties, researchers began to realize that synthetics unknown in nature had specific action that could make them useful as pest control tools. Systematic searches for such activity, with new compound synthesis to back it up, was undertaken in earnest.

"How does it work?" was a question put to me regarding 2,4-D years ago at a meeting of farmers being introduced to this—then new—herbicide. The questioner chided me a bit when I could only describe the slow abnormal growth response of treated weeds with death or recovery depending on dosage and other factors. I confessed our lack of knowledge of the biochemical processes within treated plants, but solved my problem by reminding the audience that after more than half a century of wide-scale use of aspirin, medical physiologists had only a partial understanding as to how this remarkable drug does its work.

Through diligent research, some insight into biochemical mechanisms has now been gained for a number of pest control chemicals, but in spite of incomplete knowledge we can often describe the mode of action in a general way. Many fungicides are toxic to spores landing on the leaf of a host plant, but are effective only to the extent that leaves are kept covered with the toxicant during periods of spore germination. Others, entering into plant roots or foliage and working internally after being translocated, are said to be systemic in their action. Systemic fungicides are among the most promising new types of pest control chemicals. Gardeners in a number of countries can now apply such a material to soil around roses for the control of powdery mildew and blackspot, while farmers will soon have systemic fungicides for seed treatment for the control of certain seedling diseases (Chapter 9).

Herbicides that kill foliage quickly are often called contact agents, while those that are translocated from leaves to underground parts are placed in the systemic category. Examples of the latter are the selective "hormone" type weed killer you may have used for dandelion control in your lawn and the brush killer farmers use for eliminating woody growth such as mesquite from their range and pasture land. Some herbicides, applied to the soil surface immediately after planting a crop, selectively kill germinating weed seeds and are referred to as pre-emergence herbicides. The soil-applied weed killers that give a general herbicidal effect and are residual in their action are referred to as soil-sterilants. They are used on railroad beds, parking areas and industrial grounds that are topped with gravel or cinders to prevent erosion.

Some rodenticides are acute poisons, but the widely used warfarin and related compounds control rats and mice by inducing internal bleeding over a period of days through anticoagulant activity. Insect control agents—the pyrethrins in household flying insect sprays, for example—may act quickly, while others act more slowly but result in morbidity or mortality soon enough to provide needed control.

Recently, chemical insect control agents having unusual modes of action have been discovered, and some are being made available for use. Some are systemic; they move upward in seedlings when applied with the seed or are absorbed and translocated within the plant following foliar application. Others are antifeeding agents that destroy the insects' appetite and hormones that prevent the insect from going through its normal life cycle by promoting early and lethal metamorphosis by keeping it in the larval stage, thus interfering with reproduction. There are also new sex attractants, often called pheromones, that can attract insects to traps or away from places where their presence would be objectionable. These chemicals have been particularly valuable as an aid in monitoring populations to determine when economic thresholds are reached, so that insecticide applications can be made at the most appropriate time. Some pheromones so confuse adult insects at mating time that they fail to reproduce. If these can someday be made commercially available they should prove of value in the control of a number of troublesome species.

On the horizon are new growth regulators that modify plant form in a way similar to that by which some pests are minimized or are more easily controlled. For example, there is an experimental chemical that reduces the plant height and the degree of branching of cotton. Thus, with a more open growth, insecticidal sprays penetrate the leaf canopy more readily and boll rot is less likely to do significant damage.

I recall as a boy being sent to the hardware store by my grandfather to buy some lead arsenate. The Colorado potato beetle was ruining his potato patch. It must have been about 1920. The clerk weighed out the white powder in a paper bag and then attached a sticker with a skull and crossbones on it with the word "POISON." No use directions, no precautions, and no assurance that the toxic powder would not get confused with something else. It was almost an accident that a poison warning got placed on the sack, because for a while the clerk couldn't find one. Now, only a certified farmer or custom applicator would be allowed to purchase a material with the toxicity of lead arsenate. Today the pesticides available over-the-counter to home gardeners are far less hazardous; all are labeled with clear-cut directions for use and carry appropriate precautionary statements. No pest control product can be sold except in the original sealed container. We have come a long way in insuring pesticide safety.

Once the new generation of pesticides came into wide use the old requirement of a poison label (or not, depending on the compound's acute toxicity) hardly seemed to provide adequate public protection; many countries therefore established regulatory agencies which required proof of efficacy and safety, before new products could be sold. Decisions by regulators regarding the safety of proposed applications of pesticides are now based on lengthy tests for toxicity, at varying dosage levels with laboratory animals used as indicators, and also on a series of experiments designed to determine environmental safety. (See chapter 31.) All substances sold for pest control are subject to these extensive tests, including natural products such as pyrethrins and rotenone or other plant extracts. The fact that a chemical is synthesized within a living plant—by far the world's most complex chemical factory—rather than in one designed and operated by a man has no bearing on the question of safety. Think of the many substances of plant origin that are highly toxic at adequate dosages: solanine from members of the nightshade family, and the toxin from poison hemlock that did Socrates in, for example. (See chapter 30.)

The banning of cyclamates and the threatened removal of saccharin from the market because gross amounts were found to induce cancer-like tumors in laboratory animals has made everyone aware of some of the intricacies of the U.S. food and drug law. As interpreted by regulatory agencies, these laws allow for tolerances of trace amounts of chemicals depending on the safe-level findings of toxicologists, *except* when a compound is found to induce tumors in test animals, regardless of how much the lowest dose having an effect may exceed the highest level of anticipated human exposure. (See chapter 29.) The same policy holds for pesticides, and some have been removed from the market based on indications of possible carcinogenicity at high doses in the extended tests required in recent years for older—as well as recently discovered—products.

Chapter 31 will describe current systems of pesticide regulation and the safety precautions that are being taken to assure proper use, safety to people, pets, livestock, food crops, and the environment in general, while other chapters in Part III will take up the many aspects of pesticide safety. The story of pesticides' contribution to our health, our food supply, our comfort, and to environmental protection is one that cannot be told in a few paragraphs, so read on to Part II.

As will be discussed in the next chapter, proper timing for optimum results is a key requirement for success with the chemical phases of pest management programs. If you apply a preemergence crabgrass control agent to your lawn after seedlings of this weed are up and growing, your results may be disappointing. A corn grower who sprays for cornborer before the moths have laid their eggs, which are about ready to hatch into tiny worms, will not realize a good return for his efforts. An understanding of the life-cycles of pests and their development under varying moisture and temperature conditions is an important part of modern scientific farming. Advice on application timing by specialized pest control agricultural extension workers, as well as by commercial consultants who understand all these factors and who continuously monitor populations and their development, is now available to growers of a wide range of crops.

# 8

## Putting It All Together

Not long ago, I visited a grain elevator in North Dakota where large quantities of wheat are loaded into railcars during the harvest season and where more is stored throughout the winter and spring months. Rats have traditionally been the bane of such facilities, and losses of grain from mold and insects have often been of serious proportions. When asked about current losses, the manager acted surprised at the question. He exclaimed, "We're not in Asia where they still have big losses of rice and other grains! We've got a pest management program that keeps rats and bugs down close to zero!"

When I asked how they did it, he replied that they have good structures. The old wooden ones have metal bottoms so rats cannot work their way in and, of course, the new ones of steel are relatively rat proof. Then, grain is dried when necessary before going into storage, to help prevent mold. Thermocouples are used to monitor temperatures in the grain mass and, if a "hot spot" resulting from mold is identified, the grain is turned by mechanically conveying it from one bin to another. Openings near the top of structures are screened to keep adult insects out. Rigid sanitation is practiced to keep food from rodents; spilled grain is cleaned up promptly. Finally, rodenticides are used to help keep down the rat population, and grain stored for a long period is fumigated as necessary for insect control.

Here is an excellent example of an ongoing integrated pest control program. The dairy farmer who keeps fly populations low by not allowing manure to accumulate in summer and then uses approved insecticides in his barn and milking establishment is also an integrated control person. So are you when you build your home with metal shields directly over the foundation of your frame structure and then use wood that has been impregnated with preservative. By doing this, you are helping minimize the chances of dryrot and termites. Further integration comes·with treatment of the perimeter soil with chlordane or another termite-toxic chemical that will retain its effectiveness for several yeas. The final step in your integrated control of the subterranean termite is to check occasionally for damage and for their mud tunnels built up from the ground to wooden portions of your house. A professional pest control service which provides periodic inspection can often pay off by spotting potential trouble before damage is done. And don't forget the drywood termite that occasionally infests the upper part of your home, particularly in Hawaii, California and Florida. Fumigation by a professional may be necessary if this one becomes established.

Chances are you are among the millions who practice integrated control of houseflys; screens, a flyswatter, and sanitation to avoid attracting them and encouraging reproduction together with a "pest strip" in the house and a flying-insect aerosol on hand in the event a door is left open and the population of these pesky creatures gets intolerably high. Perhaps you include sticky flypaper in your integrated program.

Although integrated pest management of certain species is hardly new, recent research has greatly broadened our horizons regarding the most effective use of pesticides with maximum opportunity for biological controls and cultural practices to play their role.

There has been much progress in determining the optimum

timing for pesticide applications. Far more than population counts are involved; the degree to which a host may be adversely affected at different stages of development is an equally important consideration. For example, the bud mite of citrus which attacks young fruit has little effect later on; a relatively low population can do great damage at one season while a high population may be harmless at another.

An insight into modern integrated pest management, often referred to as IPM, is provided by the following excerpt from an article in *Science in Agriculture*, Vol. 23, No. 1 (Fall 1975), a publication of Pennsylvania State University, by P. D. Mowrey and Dean Asquith:

A computer simulation model of the European red mite, *Panonychus ulmi*, and its apple orchard environment has been developed at the Fruit Research Laboratory. The model can simulate a mite infestation, complete with predators and environmental conditions, at the rate of a growing season per minute. This enables researchers to evaluate various pest management strategies without disrupting the real orchard. The model can be used to develop a decision-making system to predict the need for a chemical miticide.

The computer model is part of the integrated mite control program developed at the Fruit Research Laboratory and now in use throughout Pennsylvania. This program combines biological control of mites with chemical control of other pests. Biological control is achieved by the ladybird beetle, *Stethorus punctum*, a native predator of mites in Pennsylvania.

By applying reduced dosages of insecticides for other apple pests, the apple grower practicing integrated mite control encourages the predator beetle to migrate into his orchard. The beetle, in turn, feeds on the mite population and produces offspring who join in the feast. This often completely eliminates the need for a miticide. This program has reduced the amount of miticides used in Pennsylvania by 75 to 80 percent, resulting in savings to the grower and a benefit to the environment.

This work has resulted in a predictive chart which enables a grower faced with a potentially injurious mite infestation to decide when a miticide is necessary. To use the chart successfully an orchardist must make periodic counts of mites on his leaves. He can thus economize by using minimum amounts of this crop protection chemical and reduce the chances of disruption of the natural balance of predator and prey populations.

Such integrated pest management programs now being put to work in many areas are neither simple nor easy. To use them successfully, growers must be well informed and alert, and they must work hard at putting the pieces of the program together. Moreover, IPM programs developed for one area can seldom be used successfully in another without modification to fit the requirements of different species and varying environmental conditions. Example: the major natural enemy of destructive mites of apples in some areas is not a ladybird beetle as in Pennsylvania, but another mite, a predator, that attacks the plant parasitic species.

Timing of insecticide applications in modern integrated control practice depends on frequent insect counts to keep tab on population trends, just as in the Pennsylvania mite management program. With some insects, spraying is not indicated until the population reaches a damaging threshold. With those species that go through more than one generation in a season, the most effective timing may involve application to the first brood, even though populations are relatively low, so that the second brood will be less likely to reach staggering numbers.

To choose the best time for an insecticide application, one must understand the degree of damage that may be sustained at different stages of crop development and monitor populations accordingly. For example, in an integrated cotton pest management program in the San Joaquin Valley of California, it is known that lygus bug, a major component of the total insect-mite complex in that area, does its greatest damage within two weeks of the initiation of cotton flowering. Even large populations of this insect do less damage later in the season. By careful monitoring of lygus numbers beginning with the first flower formation, an appropriate insecticide can be applied at the time it does the most good. By avoiding heavy applications later in the summer, the natural predators and parasites that help keep the population of various late-season worms in check are given minimum disturbance. This program will require modification if insect species not now prevalent in the area should become a problem. Other cotton-growing areas with different insect and mite species may find the San Joaquin IPM program of little value; research must be conducted locally in order to determine which integated programs are most useful and effective in a given area.

A major objective of today's pest management researcher is to determine what various methods can contribute, and how different approaches to control can best be dovetailed. A key to sound IPM is to use pesticides when needed but in the most effective way possible. When a pesticide is required, the dosage, the best timing, and the most desirable equipment to employ must be determined; the chances must also be minimized that reducing the population of one pest will inadvertently result in the upsurge of another. This requires a rifle rather than a shotgun approach.

Much research progress has been made in recent years in the development of what I like to call "target-hitting" methods of pesticide application. Specially formulated baits for controlling cockroaches in your house or snails and slugs in your garden are well-known examples. Another you are no doubt familiar with is the flea collar on your dog. Instead of spraying or bathing your dog with an insecticide, you can employ the flea-toxic substance impregnated into a plastic collar. This keeps fleas in check—and some ticks, too—with very minimum loss of the insecticide down the drain or into the atmosphere.

Cattle are now provided with backrubbers containing insecticides toxic to lice, in lieu of the old spray method of control. An annual systemic cattle grub control treatment is applied as a dipperful of an emulsion of a dermally-absorbed insecticide poured on the animal's back rather than the old drenching spray. Obviously, the compounds used have a low level of toxicity to warm-blooded animals.

Livestock people are not the only ones who can now save money and do a safer and better job of pest control through target-hitting techniques. Bananas have their special treatment; a polyethylene bag with an insecticide impregnated in the plastic is placed around each bunch as it matures. With this enclosure, better quality bananas are assured and the insecticide prevents aphids or other insects from attacking the bunch. Soil insects that attack seeds or young plants are often controlled by banding an insecticide over the row, and of course treating the seed itself to protect against maggots or organisms of decay is the ultimate in target-hitting. Incorporating mildew inhibitors with paint and impregnating wood with a chemical that prevents decay and insect attack are other rifle approaches to pest control.

More familiar to householders are insecticide-impregnated "strips" designed to hang in locations where flying insects will receive a lethal dose. Recently introduced is a pressure-sensitive tape impregnated with an attractant and an insecticide. When stuck to the floor in dark places frequented by cockroaches, i.e., under sink, stove or refrigerator, this pest can be kept in check effectively. A recent target-hitting innovation is a microencapsulation process for a cockroach insecticide that enhances safety and assures long-lasting control.

Now chemists have devised slow-release formulations for specialized uses such as ant control, where the toxicant must be carried back to underground nests for the greatest effect. Formulation chemists have also devised a granular insecticide for mosquito control that gradually releases a larvacidal concentration of a specific compound into the water of breeding sites over a whole season without reaching a concentration harmful to many forms of wildlife.

Spraydrift is minimized by employing equipment and formulations that provide droplet-size control and thus reduce the chances of atmospheric contamination. Regulating both droplet size and the wetting properties of a spray are important to maximizing retention of a pesticide on the sprayed surface.

Adjuvants that improve selectivity as well as ones that increase activity have been discovered and put to use, thus minimizing the amount of toxicant required for a given degree of control. New application technology is making it possible to recirculate spray that does not contact the plant by collecting unimpinged droplets in a "catch basin" provided as an integral part of the equipment. Electric-eye controlled sprayers now deliver a pesticide only when the nozzle is over the plant; it cuts off delivery between plants. Thus, chemical is saved and the chances of atmospheric contamination are reduced.

Tall-growing weeds are sometimes treated by a device that rubs off a systemic herbicide on the parts of the unwanted plant extending above a crop, the "wiper" type of applicator. You may have used a modification of this scheme in the form of an herbicide-impregnated wax bar which, when pulled across your lawn, rubs off on foliage of grass and weeds alike, but only the weeds are affected, because of the selective nature of the chemical.

The nature, the extent, and the cost of potential damage from a pest must all be considered in making pesticide use decisions. For example, a fungicide treatment for the control of leaf diseases may not pay off for wheat, and a grower will have to accept the best yield he can obtain with only genetic or cultural controls. For more valuable crops such as celery, potatoes, and bananas, an integrated disease control program that includes a fungicide may be imperative if the producer is to survive economically and if you are to find his products in the marketplace.

A worm hole in an edible plant part must be taken more seriously than a hole in a leaf whose function is photosynthesis, but if too many holes are chewed in leaves, the whole plant growth system may fall far short of its potential. For example, a relatively few adult codling moths laying their eggs on apples can result in intolerable fruit losses. A single worm hole makes an apple unacceptable for market and it would not keep well in storage even if customers were inclined to accept insect-damaged fruit. On the other hand, it takes a sizable population of leaf-eating canker worms on forest or street trees to do appreciable damage. Obviously, an insecticide must be used more frequently for codling moth control than for canker worm, but in either case, sound management requires frequent checks of populations to determine when an application of an appropriate chemical is necessary.

Because of the importance of moisture droplets to the germination of fungus spores, proper timing for foliar fungicide application often depends on accurate weather forecasts, as well as knowledge of the presence of inoculum. As forecasting improves, fruit and vegetable growers will be able to do a more effective job, sometimes with fewer fungicide applications. With the wider use of satellite photographs (remote sensing), the spread of wind-carried fungus diseases can be plotted as an aid in predicting when treatments should be initiated. New computerized services are springing up which make it possible for a specialist at a central location to make accurate disease forecasts and for growers to determine by telephone just when a fungicidal spray should be applied for greatest payoff. It must be recognized, however, that different fungi have different meteorological requirements for optimum development, so when two or more diseases are prevalent—e.g., early blight, late blight, and Botrytus blight of potatoes—the number of fungicide applications based on the best of multiple-disease fore-casting methods may be no fewer than when an old empirical spray schedule is followed.

With the research effort now going into improved inte-grated methods of pest management, better control with an adequate margin of environmental and human safety should be available in the future. Much depends, however, on the dis-covery and subsequent commercial development of safer and more effective specific pesticides for the chemical phases of in-tegration. Unfortunately, the rigid governmental regulation of pesticides in recent years has tended to discourage investment in the research so essential to our future food supply, our health, and our economic well-being. More on what we must do to assure better pesticides for integrated pest management programs for the future will be discussed in the final pages, under the heading "Trade-Offs."

Recently several writers have equated IPM with biological or other non-chemical methods of control. The very term, inte-grated pest management, implies the utilization of whatever methods are available in one concerted program aimed at avoiding undue losses. True, a few pests can be effectively man-aged by integrating genetic, biological, and cultural methods alone, but they are in the vast minority. Chemicals still pro-vide an essential input to most successful IPM programs. Un-fortunately, some writers tend to oversimplify the task of de-fending ourselves against pests, and they often present research progress on IPM methods as if these were proven programs ready for farmers and gardeners to apply. Pest management programs should be shown to be effective in carefully con-ducted and repeated field tests before being offered to the public as proven methods of controlling a given pest.

# PART

# II

## *The Pesticide Drama*

[Some scenes from modern pest control technology in action]

# 9

## It Begins with the Seed

The old fieldhand helping me plant sweetcorn variety trials at the Alabama Experiment Station chanted as he counted out four seeds for each hill:

> "One fur de cutworm,
> One fur de crow,
> One tu damp off
> One tu grow."

That was over forty years ago. If his verse were still apropos, the United States could not possibly have produced more than seven billion bushels of corn in 1979. Stands would have been reduced on many farms as a result of seed decay or insect damage. A large number of fields would have required replanting at a considerable cost in seed, manpower, and tractor fuel. Most soils are teeming with organisms of decay which can cause stand reduction if germination is greatly delayed because of cool weather. Seed corn maggots and other insects are often ready to devour the kernels as soon as they are in the ground.

Few of these bad things happened to newly planted corn seed during the last two or three decades because of the almost universal use of seed protectants. We have no successful biological or cultural control approaches to seed decay or seed attack by insects, but we do have effective chemical treatments. Because decay-preventing fungicides are applied by the commercial seed producer as a film of chemical on the seed coat, the farmer often does not realize how important this phase of crop protection is to his success.

Early planting cannot be carried out with confidence without seed protectants. Too often, a farmer would have a reduced stand or would have to replant because of seed decay associated with cold soil. Early planting permits the use of higher yielding long-season hybrids and/or higher populations of early hybrids and enables the crop to take advantage of the favorable moisture conditions usually prevailing during the first part of the growing season. Further, early plantings utilize soil nitrogen more efficiently. These factors account for at least 10 bushels per acre or more than 700 million bushels annually in the United States alone.

Without seed protectants, even with the old later date of planting, there would often be weak stands. Today's corn grower recognizes the importance of high plant populations to top yields, and to assure them it is important that he be able to plant the right number of seeds with confidence that most will come up. This precision is impossible without seed protectants. These chemicals have contributed at least another five bushels per acre to U.S. average yields by permitting precision planting and assuring optimum stands.

Assuming 70 million acres of corn and a price of $3.00 per bushel, we can add up the cost resulting from *not using* chemical seed protectants for this crop in the United States.

(1) Yield loss if we return to the
    later planting of short-season
    (early maturing) hybrids.
        700 million bushels @ $3.00 . . . . . . $2100 million
(2) Yield loss from less-than-optimum
    stands that would be inevitable
    without seed protectants.

350 million bushels @ $3.00 ...... $1050 million
                                    Total <u>$3150 million</u>

To put it another way, the United States realizes—aside from costs of occasional replanting—more than a $3 billion-a-year gain in corn productivity because of fungicidal seed protectants (which cost an average of less than a dollar for enough seed for an acre).

Aside from the importance of that crop to domestic supplies of meat, milk, and eggs, as well as cereal products, an additional two-billion-plus bushels of corn has a dual benefit: to an improved U.S. trade balance and also to the adequacy of the diet of importing countries. In the future, some corn will be used for fermentation to produce alcohol to blend with gasoline to make gasohol for motor fuel.

Perhaps the greatest value of seed protectants is the stability they contribute to production. Predictability of crop yields is important to planning by individual farmers as well as by governments and industries utilizing farm commodities. Improved certainty of production is a vital factor in assuring an adequate diet for the world's more than four billion people. Conversely, unexpected dips in production such as might be caused by adverse weather following planting without the benefit of seed protectants could trigger increased inflation. Even now, replanting is occasionally necessary, but it would be far more common without protectants. Aside from added costs for seed, labor, and motor fuel, extra trips over the field with a tractor and planter would contribute to soil compaction, a major environmental concern of those who produce your food.

An appreciable percentage of the seed used for planting wheat, oats, barley, and rice is also treated with a protectant to insure good stands. With these crops, there is seldom an opportunity for replanting; getting a stand adequate to provide an optimum yield from the first seeding is therefore very important. Cotton seed and grain sorghum, too, are usually treated with a protectant to prevent decay. Many vegetable seeds are automatically treated with a fungicide by the seedsman, so a grower may not recognize the benefits received through better stands. Potato tubers used for seed are very subject to decay, and with a planting rate of many hundredweight per acre, a fungicidal and bactericidal treatment to prevent loss is highly essential to an assured stand and profitable production.

Farmers sometimes apply an insecticide with seed of crops that may be damaged before they emerge, and some seed is pretreated by the supplier for protection against maggots and other underground insect pests. Several systemic insecticides applied to the seed will be taken up by young plants and distributed throughout the stems and leaves. Thus they can give protection against certain seedling insects, for example thrips of cotton. An outstanding recent advance in agricultural technology is the discovery of systemic fungicides that protect seedlings from disease. For example, a summary of progress at the International Rice Research Institute in the Philippines for 1979 states: "Several systemic fungicides applied as seed treatments at low rates gave good control of leaf blast for as long as six weeks. The seed treatments were easy and economical."

# 10

## Famine on the Wind

Those early Spanish explorers who took tubers of that strange Andean plant, the potato, with them on their return from the New World would have been astonished to learn how important the crop had become, only a few generations later, as a source of food throughout much of Europe. Then emigrants carried it from Europe to the far corners of the earth, including North America. In cooler climates it soon became a food staple.

Actually some people became too dependent on the potato, as you will recall from history of the Irish famine of the late 1840s. When the then uncontrollable late-blight disease caused by the wind-blown fungus *Phytophera infestans* virtually destroyed the crop, nearly a third of the population of Ireland died of starvation, and another third was forced to migrate—largely to the New World.

When one considers the importance the potato achieved in spite of all the troubles the crop has experienced, it is quite evident that it is an efficient producer of food, and one that people like to eat. Seed-piece decay after planting; potato beetles, leaf hoppers, and assorted other insects attack the plant; there is early-blight in some seasons and late-blight in others, and scab on the tubers; there are various wilts to cut the crop short; and following all this, there is spoilage in storage. With all these problems—few of which could be controlled during the first

300 years of the potato's European and North American domestication—why did people try to raise this crop?

Even as late as 1940 potato pest control was not very efficient, and yields per acre were low compared with those of the present. Let's look at a 60-year yield history in the United States.

| Decade beginning | Per acre yield of potatoes in hundredweight |
|---|---|
| 1920 | 56 |
| 1930 | 67 |
| 1940 | 82 |
| 1950 | 165 |
| 1960 | 208 |
| 1970 | 247 |

Note the marked increase in the decade of the 1950s compared with the yield 10 years before. What brought this about? It couldn't just have been better varieties, because some of the older types grown for decades were still planted and showed a corresponding increase. No doubt some yield enhancement was due to improved plant nutrition, but potatoes had usually been well fertilized with manure and mineral supplements. Furthermore, they were often grown in rotations with legumes. A major factor in this sharp yield increase: *markedly improved pest control through the use of new chemicals.*

I never knew what a healthy potato leaf looked like until the late 1940s. I thought their leaves were naturally cupped or rolled downward, often tending to burn at the edge. Then, when we finally had an insecticide that controlled leaf hoppers—and thereby the die-back of leaf margins called hopper burn—I learned that normal insect-free potato leaves were large and flat. Without hoppers the vines grew more vigorously than I had ever thought possible, and with greater leaf surface and thus increased photosynthesis they could naturally produce more and larger tubers.

For many decades the old Bordeaux Mixture discovered in France during the latter part of the nineteenth century was widely used on potatoes for blight control but, in itself, it had a phytotoxic effect. There was often a net gain in yield resulting from fungus control, but how much better it would have been if an equally effective material with a less drastic effect on the potato plant had been available. Such compounds actually came along in the 1940s in the form of the dithiocarbamate fungicides, still widely used for blight control.

There are many other types of pesticides useful for potatoes; for example, tuber-treatment compounds to prevent decay after planting and insecticides for the control of flea beetles, potato beetles and aphids. Nematodes can be troublesome and, where rotations are inadequate for their control, nematicides are used. Chemical vine-killers which stop growth quickly when the crop is "made" are often useful for reducing the spread of the late-blight fungus to the tubers and thereby help prevent the storage decay associated with this organism. Soil fumigants help minimize verticilium wilt in some important potato-growing areas.

If we were to go back to the potato culture of the nineteenth century, I doubt that the yields would average one-fourth what they do today. No doubt productivity would gyrate wildly, with some years being almost a complete loss and others showing reasonably good crops. If we went back to the days of my boyhood potato-growing experience of the 1920s, we'd have only an arsenical spray for beetles and other chewing insects, nothing at all for leaf hoppers, and only Bordeaux Mixture for blight. I doubt that any of the 1920 potatoes that I helped pick up on nearby farms produced over 75 hundredweight per acre; in fact, I remember a neighboring farmer boasting about raising over 100 bushels per acre—and this, of course, would have been only about 60 hundredweight. Note again the yield trends of the past half century and the average productivity during the most recent decade of 247 hundredweight or more than 400 bushels per acre. Similar yield increases have occurred in all of the important potato-growing countries. Some of this progress is due to better varieties and to improved fertilization, but improved crop protection chemicals are prime factors.

There is currently much research emphasis on breeding potatoes for resistance to various diseases, and the biological control of certain insect pests of this crop holds much promise. Hopefully this effort will prove productive; however, at present—and, I predict, through the rest of this century—pest control chemicals will form the backbone of any integrated pest management system for this important food crop.

# 11

## Rotten Wood and Tumbling Buildings

If you have ever tried to make your way through an old forest, you know what an obstacle course fallen trees and branches can make of the forest floor. Now visualize what an impossible tangle there would be if past generations of trees had not been reduced to surface litter by insects and decay organisms. Fortunately nature provided termites, ants, beetles, and a myriad of cellulose-rotting organisms that keep our fields and forests from becoming nature's junk heap.

Almost everything of organic origin is biodegradable. If it weren't, the ground would now be so shaded by debris that few plants could grow. But we don't want everything we have produced to degrade! Certainly not our stored food nor our wool and cotton clothing. Neither do we want railroad ties, electric poles, fence posts, and wooden structures to rot away after only a few years. They represent valuable forests and much labor.

Long ago we learned that wood kept dry will not readily decay. Termites and other insects may still attack it, but dry wood doesn't ordinarily rot. But how do you keep a post or pole or railroad tie from being moist? For that matter, how do you keep wooden structures in warm humid climates dry enough to avoid decay? The answer to this problem has been found in wood preservatives that not only prevent decay, even with constant dampness, but also repel attack by termites, powder post beetles, and other wood-eating insects. Conserving things we have grown is just as important to wise land use as any other aspect of conservation. If we have concern for forest protection and think that more scenic areas should be set aside for recreational purposes, wood preservatives must be considered among our most important groups of pesticides; widely used but seldom fully appreciated, they are truly conservation chemicals.

Most telephone and electric utility poles, railroad ties, and dock pilings are now impregnated with pentachlorophenol, creosote, or some other wood preservative. Fence posts and structural lumber are increasingly being treated prior to use, particularly where both decay and termites cause severe losses. It is estimated that over one billion cubic feet of wood are treated annually in the world. Assuming that preservatives make wood, in decay-prone situations, last an average of 50 years compared with 10 years if untreated, we can say that five times as much timber would have to be cut from such areas annually, if preservatives were not used. At average growth increments now realized in well-managed forests of the northeastern United States, it would require a forest twice the size of all of New England to produce an extra four billion cubic feet of wood annually on a sustained basis. Surely wood preservative technology has made a valuable contribution to a better environment.

About ten years after the close of World War II, I was on a decommissioned airbase in the South where barracks had been abandoned. Subterranean termites had riddled the buildings from below, and the drywood type was swarming over what was left of sagging rafters and roof boards. Walls had collapsed, and those buildings that had not already tumbled into a tangle of rotten lumber were certain to do so with the first strong wind.

The experience made me realize how important our defense against termites really is. In termite country, our homes and other wooden buildings have been protected for so many years we have all but forgotten how much damage these insects can cause. Perimeter treatment with a residual insecticide for the subterranean termite is an important defensive weapon which, along with shielding against access from the soil, enables us to keep wooden structures sound almost indefinitely. Fumigation for the drywood termite contributes additional savings in wood and a corresponding reduction in the area of timber that must be cut each year, not to mention the protection of people's homes, places of business, and public buildings.

Preservative chemicals prolong the life of many other non-food materials that man has produced and thus contribute to conservation of resources—for example, paints, diesel fuel, and cutting oils. Algacides and slime-inhibiting compounds in cooling water materially prolongs the life of mechanical equipment. Damage from clothes moths is rare today, partly because synthetic fibers are not attacked but also because insecticides are available for mothproofing of wool fabrics and prevention of infestations in woolen mills, warehouses, and private homes.

# 12

## Thief Underground

Dr. G. Steiner, who for many years was the chief nematologist for the U.S. Department of Agriculture, told a story of a visit to home gardens in the South while scouting for various parasitic nematode species. He spotted a sickly row of tomato plants in one garden and guessed that they were all suffering from rootknot, a common ailment of crop and garden plants throughout the warmer regions of the earth. After asking the owner's permission, Steiner dug up one of the plants and, as he suspected, only grossly malformed nematode-infested roots were struggling to keep the above-ground plant alive. When he remarked about the severe infection the owner said, "Why, that's the way tomatoes grow!" Apparently the home gardener had never seen a healthy tomato plant or normal tomato roots.

Rootknot is so common that there are millions of people who really don't know what the roots of their beans and okra and cabbage and many other crops would look like without it or how productive these crops might be if not infected by nematodes of the genus *Meliodogyne*, the rootknot culprit. There are many other nematode species acting as underground thieves and attacking a wide range of crops.

Several natural enemies of nematodes—including certain soil fungi and predaceous nematode species—help keep populations in check, and these are encouraged by organic matter in the soil. Rotation is helpful, but even the cover crops and weeds that occupy land not being productively farmed can harbor many species of these very tiny worms that crawl into roots to feed or attach themselves, leach-like, to their surface and extract nourishment from the host plant. Where these natural and cultural controls are inadequate, normal root development is prevented. Without a healthy root system there is no chance for a highly productive crop.

In recent years, research has provided us with specific soil-applied chemicals to reduce the population of nematodes before planting. In many situations these have given a big boost to the control provided by rotation; crop yields are often increased by tons per acre as the result of their use. Practically all pineapple is treated with a nematicidal fumigant. Costs of producing this crop are so high that no grower can afford the fertilizer and land preparation and the extensive labor for planting unless he obtains a very good yield. With rotation alone as a control method, and without nematicides, pineapple production in much of the world would be quite inefficient. If this favorite fruit was still grown with the inevitable low yield caused by various species of nematodes, costs would be so high that only the rich would be able to afford it. Nematode control is one of the factors that has made pineapple a common processed fruit on our supermarket shelves and, in many areas, an economical fresh fruit as well.

Soybeans, now grown on more than seventy million acres in the United States, are increasingly attacked by the cyst nematode and, in some areas, by other types. Considerable progress has been made in breeding soybeans for resistance to nematodes, but when populations of this parasite are especially high, help is still needed from a nematicide. Recent research by R. A. Kinlock of the University of Florida showed that the soybean variety Bragg, which is resistant to rootknot, was still increased in yield by one-third through the use of nematicides when grown on heavily infested land. In these tests Hood, a susceptible variety, produced next to nothing without a nematicide. Here are Kinlock's data:

| Nematicide | Yield in bushels* per acre | |
| | Hood (susceptible) | Bragg (resistant) |
| --- | --- | --- |
| None (untreated control) | 3.1 | 24.6 |
| A | 25.1 | 38.0 |
| B | 20.4 | 32.6 |
| C | 17.4 | 38.9 |

*One bushel of soybeans weighs 60 pounds (27.2 kilos).

As soybean production increases, various species of nematodes are likely to build up in some soils. Only by an integrated attack including rotations, the maintenance of good soil fertility, the use of resistant varieties and on some soils treatment with nematicides can optimum yields be obtained.

Let's look at sugarbeet production in relation to the prevalence of the nematode *Heterodera schachtii*. Where land is available for long rotations, the population of this soil pest does not build up to highly damaging proportions. In the extensive beet-growing areas of Europe and eastern North America, rotations of 7 years are often needed to keep the pest in check. But in many productive intermountain and coastal sugarbeet-growing areas of the western United States, there simply is not enough land to permit control by rotation alone. Dichloropropene soil fumigant has made it possible to produce large acreages of beets economically on land that would otherwise have to be turned over to some other crop. Without fumigation, yields on infested land would not exceed ten tons per acre while, with fumigation, twenty-five tons or more are common.

We all saw how precarious is the balance between supply and demand of sugar when sugar prices skyrocketed in the autumn of 1974. Western United States sugarbeet areas are important to world supply of this commodity and, were it not for nematicides, their acreage and productivity per acre would be much lower than it is today. Without the contribution of these specialized pesticides we would very likely again have a sugar shortage and wildly-gyrating prices.

The thief underground will continue to be there. Nematode-resistant varieties, when available, and rotations, to the extent they are practical, play an important part in nematode control. But nematicides—of both the soil fumigant type and the newer nonfumigant compounds—contribute significantly in providing us with additional supplies of such widely diverse crops as bananas, beans, lettuce, cotton, carrots, strawberries, pineapple, potatoes, tomatoes, beets, oranges, peaches, grapes, coffee, tea, mushrooms, and the cruciferous vegetables.

# 13

## How Much Grass for Lunch?

The nine students I took on a field trip to visit some Michigan farms were ravenous after a long morning of driving and tramping, so each ordered two giant hamburgers at lunch and most drank two glasses of milk. This was topped off with double-deck ice cream cones. When it was all downed, I suggested they figure out how much converted forage the ten of us had consumed—five pounds of hamburger, more than a gallon of milk and at least two quarts of ice cream. Assuming the beef and dairy animals that provided this food were raised largely on forage, it added up to about 250 pounds of grass, clover or silage.

Much has appeared in the press recently regarding the inefficiency of the beef animal in converting grain to meat, but it is often overlooked that most of the animal's weight is from converted forage. Even high-grade beef animals are finished on grain for only the latter part of their life. Many receive only grass, particularly the grades that are used for processed meats and hamburger. Dairy animals, likewise, get much of their nourishment from forage of one kind or another. The forage consumed by the world's billion or more cattle is truly tremendous, exceeding by far the tonnage of all food consumed by people.

The major ruminant animals—cattle, sheep, and goats—constitute a blessing or a curse for mankind, depending on grazing practices. Historically, vast areas have been all but denuded of soil because too many animals have been allowed to graze the vegetative cover too closely—the Mediterranean region, for example. Often the demand for wood for construction and fuel prompted deforestation, and then overgrazing prevented re-establishment of either trees or stable permanent pasture vegetation. Goats and sheep were particularly involved in inducing erosion, because they bite forage off closer to the ground than do cattle. Even today, in spite of our greater knowledge of sound grazing practices and our capability of improving pastures with better-adapted forage plants and of enhancing the density of growth through fertilization with those minerals a particular soil lacks, there is still much land degradation as a result of destructive grazing. Floods downstream from over-farmed or over-grazed foothills of the Himalayas and desert encroachment south of the Sahara are prime examples.

On the other side of the coin, the ruminant animal can provide man, on a perpetual basis, with vast quantities of needed protein as well as shoes and clothing, if pastures and range land are not stocked beyond their carrying capacity. Much of the world's land surface is ill-suited to crop production because of arid climate, excessive slope, poor drainage or lack of fertility. Obviously much of this should be left in forest, but much can be wisely and effectively used to produce meat, leather, wool, and dairy products. Well-managed grazing land balanced with crop production on suitable soils, with vast areas left in forests, constitutes an agricultural and environmental ideal, achievable with our present knowledge and resources if population pressures are not excessive.

In some parts of the world, grasslands will remain as such without encroachment by woody scrub, but much of our potential grazing land is gradually invaded by unpalatable bushes and trees if control measures are not taken. Some of this results from over-grazing; weakened grass cannot keep out encroaching woody vegetation. But much is ecologically inevitable, particularly in regions that supported forests or scrub growth before man entered the picture. If you live in such an area you may have noticed how vacant lots or abandoned fields are overrun with woody plants after a few years. The brush-infested grazing land one sees in so many places is not producing livestock efficiently, because woody plants shade the grass and compete for nutrients and water.

From the standpoint of soil and water conservation and efficiency of utilization, grass often provides the optimum cover. In Oklahoma measurements made during an eight-year period showed that 45 percent less water ran off annually from good grass on brush-free herbicide-treated land than from an adjacent area of brushy pastures. There was also less soil erosion on the treated area. In these tests, surprisingly large yields of nutritious forage was realized following use of selective herbicides.

Brush has a much deeper root system than grass and draws water from lower depths, thus exhausting the subsoil reserves that during dry weather move upward to the zone of grass root growth through capillary action. The subsoil, often robbed of moisture by brush, is also important to the stock raiser as a reservoir for feeding the streams and wells he must depend on for watering his animals. Many ranchers in semiarid areas have found that wells yield water and streams continue to flow much longer into periods of drought, where woody plants have been controlled.

Under ecological conditions favorable to woody growth, it takes diligence to keep nutritious grass from being taken over by scrub. Keeping nonnutritious, unpalatable, and sometimes poisonous plants out of range and pasture land has always been a difficult task. Cutting brush is like going to the barber shop. It grows right back and has to be cut again. Bulldozing or chaining may have a temporary beneficial effect, but the cost in dollars and energy is high, and there is risk of erosion.

About three decades ago, some remarkable new chemical

tools became available to the farmer and rancher for pasture and range improvement: selective herbicides that would control woody plants without hurting grass. These have revolutionized pasture management not only in the United States, but also in Australia, Latin America, and some parts of Africa. They are, of course, not a substitute for sound grazing practices; over-grazing must always be avoided if a desirable composition of forage is to be maintained. But they help do a job that we have always known was essential but often almost impossible—the control of nonnutritious and unpalatable woody plants.

Let's look at the results of a few experiments on encouraging the more desirable components of mixed vegetation through the use of selective herbicides. Dr. Larry Mitich, of the North Dakota Experiment Station, measured the grass and nongrass vegetative growth in a pasture improvement program including plots in twenty-two counties in his state. After two annual applications of about one pound per acre of 2,4-D (one kilogram per hectare), the vegetative cover changed from 67 percent to 93 percent grass with a corresponding decrease in nongrasses, mostly unpalatable weeds. The dry weight of all vegetation after two years of spraying was increased by 60 percent compared with the unsprayed plots. Thus, the treatment with 2,4-D resulted in an increase in total growth, as well as in the percentage of useful herbage.

At the Cualaca Experiment Station in Panama, in a zone of annual rainfall of 229 centimeters (90 inches), the production of grass and nongrass vegetation in a brush-infested pasture of Jaragua grass was measured after treatment with a mixture of picloram and 2,4-D herbicides. Cattle were withheld and clippings were made 460 days after application. Note in the following table the shift from brush to grass in the treated areas, with the total weight of vegetation remaining relatively constant.

| | Results of treatment with Picloram plus 2, 4-D | | | |
|---|---|---|---|---|
| | Weight in tons* per acre** of: | | | |
| | Jaragua grass | Other grasses | Live brush | Total vegetation |
| Untreated check | 1.80 | .28 | 12.65 | 14.73 |
| Herbicide treated | 13.55 | .29 | .50 | 14.34 |

*One ton per acre equals 2.2 metric tons per hectare.
**One acre equals .405 hectares.

Today herbicides are being used to improve existing permanent pastures without the expense and risk of erosion associated with plowing up the old sod. A good supply of soil nitrogen is essential to productive forage development. With recent increases in the cost of nitrogen fertilizer and prospects of future energy shortages, it becomes increasingly important to keep nitrogen-fixing clover or other forage legumes growing in mixture with grass. Now grass-inhibiting herbicides may be used to suppress growth while legumes, seeded in slits in the sod, get off to a good start. When the grass recovers from the controlled dose of a suppressant, a greatly improved herbage composition results and good growth will ensue without the expense of nitrogen fertilization. Of course this is not a substitute for the application of phosphate, potassium or other minerals that may be needed.

Abortion of cattle can be a significant factor in the cost of producing beef and dairy products, particularly when calves suitable for herd replacement are lost. Wisconsin has been a major dairy area for a long time, and I doubt there are more efficient farmers anywhere than there are in that state which provides so much milk and cream for the fresh market and also for the manufacturing of cheese, butter, ice cream, yogurt, and the dried milk used by bakers and other food processors. Yet some good farmers had a continuing problem of cattle abortions, and a better understanding of the cause became an objective of veterinarians and agricultural experiment station livestock specialists.

It was observed that farms experiencing a disproportionate number of cattle abortions had considerable pasture on moist peat land which, in addition to grass, harbored many kinds of broad-leaved weeds. The veterinary sleuths determined that many of these weeds, consumed in abundance by livestock, tended to be high in nitrates. Earlier work, confirmed by this research, established that high dietary nitrates could induce abortions in cattle without reaching levels toxic to the mother cows. Field experiments showed that weed control in these peat-land pastures with 2,4-D was a practical way of preventing the excessive incidence of cattle abortions.

Many kinds of plants are directly toxic to grazing animals. Sometimes these hazardous species are consumed because there is a scarcity of more nutritious forage, but at other times harmful plants appear quite palatable. Nature often "knows best," but in these cases a little assistance in the form of poisonous plant control can help nature along and conserve resources by making livestock production more efficient.

Bracken fern, wild larkspur, halogeton, wild cherry, loco-weed, deadly nightshade—these are some of the species that can be toxic at adequate consumption levels. The book *Poisonous Plants of the United States and Canada*\* lists several dozen that that can be toxic to livestock. When animal deaths occur, farmers and the public are greatly concerned, and efforts are made to alter grazing practices or to control the incriminated plant species. Veterinarians believe that, aside from mortality, great economic losses occur through reduced weight gains and general morbidity resulting from sublethal intake. Some winter deaths of range animals are believed to result from lowered

---

\*John M. Kingsbury, *Poisonous Plants of the United States and Canada* (Englewood Cliffs, New Jersey: Prentice-Hall, Inc., 1964).

vigor as a consequence of subacute poisoning from toxic plants. Herbicides have proven to be useful tools for the management of a number of poisonous range and pasture species and—with environmentally and economically sound usage—can be even more beneficial in the future.

# 14

## An Apple a Day

We have a lone apple tree at our summer place in Michigan which I never spray for insects or diseases; I just never seem to be around at the right time. In three or four of the thirty years we've owned it, this tree has had clean fruit. Sometimes the apples have only modest scab and insect damage, but at least half the time they are badly riddled with maggots or coddling moth. Apparently, in some years, natural predators keep the insects in check. Other years the predators are not present in adequate numbers and sometimes the weather favors scab development. If this were an orchard instead of a lone tree, I would have long since gone broke, and the apple trees would have found their way to someone's fireplace.

If everyone raised apples in this way, there would be occasional years with a decent crop, but most of the time supply would fall so far short of demand that the prices would be astronomical. Apples and other fruits, including citrus and bananas, are plentiful in our markets in part because we have protective chemicals that control insects, diseases, and other pests. Eliminate these pesticides and good fruit would become a luxury which only the rich could afford. Even if the public would accept scabby or wormy fruit, much of it would not keep well in storage and therefore would be available through only a short season following harvest.

Not all pests of apples have as drastic effect on the fruit as the scab fungus, coddling moth, and maggots, but others can drastically reduce yield, and their insidious nature can be misleading. Mites, those tiny spider-related pests one can hardly see but which are so often present on the underside of leaves, seem to love foliage of apples and other deciduous fruits. In adequate numbers, they reduce the photosynthetic efficiency of the leaves and this, of course, shows up in greatly reduced yields. An indication of the seriousness of mite infestations may be gained from data from the Pennsylvania Agricultural Experiment Station. The following table gives the average yield in crates of apples per acre (50 trees) from trees given no treatment compared with those on which mites were controlled.

| | Yield in crates per acre from: | |
| | Trees | |
| Variety | Mite-injured trees | where mites were controlled |
| --- | --- | --- |
| Red York | 825 | 2,090 |
| Golden Delicious | 575 | 1,570 |
| Stayman | 1,000 | 1,605 |
| Rome Beauty | 1,315 | 1,770 |

Research has provided fruit growers with successful integrated systems of mite control, an important phase of which is the use of some excellent miticides. One is so active that only two ounces per hundred gallons of spray is required and so selective that, when properly applied, no adverse effect is experienced on beneficial insects or even on the desirable predatory mite species. The integration of these miticides with non-chemical control methods to form a pest management system is discussed in Chapter 8.

Healthy leaves are always essential to a highly productive crop, and the banana plant is no exception. A wind-borne fungus causing a leaf spot known as the Sigatoka disease is a serious menace to banana plantations whenever the weather is damp, which means much of the year in many highly productive growing areas of Central America, Ecuador, and elsewhere in the tropics. It was found long ago that Bordeaux Mixture, a composition of lime and copper sulfate, would reduce the severity of Sigatoka, but in more recent years the dithiocarbamates which are more effective and less likely to injure the crop have been widely used. Other fungicides have also found their place. Today every profitable and productive bananagrowing enterprise in the humid tropics depends on fungicides to control this leaf spot disease. Without such measures, yields would be drastically reduced and bananas would either no longer be on the market or would be so expensive that few of us could afford them.

In addition to bananas and apples, citrus constitutes another major fruit that is produced very efficiently and in large volume. Fortunately, citrus is not always beset with insects, nematodes, and fungus diseases, but when it is, pesticides are often needed for control. Some valuable biological controls have been developed and are operational, particularly for scale insects, but after these are used and the best cultural practices employed, there are still times when pesticides are essential if production is to be constant and the growers are to make a profit adequate to keep them in business. Obviously citrus producers must be successful if you are to have orange juice and grapefruit at a price you can afford.

An important aspect of providing you with fresh citrus at your supermarket is the prevention of diseases that can attack the fruit after harvest—for example, blue mold of oranges. You have probably seen, on occasion, a moldy orange and perhaps have noticed how the infection tends to spread from the original diseased fruit to others that contact it. This post-harvest decay is largely prevented these days by mold inhibitors often applied as a dip at the packing plant.

No aspect of conserving our resources and protecting our environment is more important than preventing spoilage of things that we have already grown. Were we to stop using conservation chemicals, all of which are approved by food and drug authorities as being safe, we would run into such tremendous losses that many fruit-growing, shipping, and processing enterprises would be forced out of business.

The old saying, an apple a day keeps the doctor away, is not pure myth. Nutritionists everywhere recommend fruits as healthful food. Time and again one reads the admonition by health authorities to avoid excessive intake of meats, sweets, and starchy foods and to substitute more fruits and vegetables. If we fail to adequately control insects, mites, nematodes, and various diseases of these crops, they could indeed become luxuries and our good health would be jeopardized.

# 15

## Chicken Any Day

While shopping for groceries one recent Saturday, I noted that the price of cut-up chicken was $0.79 per pound while chuck roast was $1.89. This set me thinking about the days I remember when chicken was more expensive than beef. Of course, all prices were lower back in the 1920s and 1930s, but poultry was then considered a luxury and always sold for more than the medium-priced cuts of beef and pork. If you happen to have read that old best seller, *Chicken Every Sunday,* you learned that this luxury was indeed reserved for a special day and chicken during the week was a rare experience.

What happened to reverse this price relationship between red meat and poultry? It was not creeping inefficiencies in the livestock industry; there have been improvements in feed conversion, disease control, and other aspects of producing both beef and pork. But why hasn't poultry increased in price percentagewise as much as other foods? The answer is that there have been some remarkable improvements in efficiency which account for the ability of the poultry industry to provide us with their product at a much lower price relative to the economy as a whole than was possible thirty or more years ago. Now we can have chicken any day of the week, and often on several days. Poultry has become one of our more affordable sources of animal protein.

What brought about these economic improvements? Some are related to processing and wholesale and retail distribution, but a great deal of credit goes to advances in production technology, all resulting from research and development involving:

1. *Better Breeding* — If you could compare the appearance of a modern dressed broiler with the kind of meat birds we raised years ago, you would see that the breeders have developed birds with larger breasts and more meat in relation to the weight of the entire carcass. They have also bred birds that have better appetites, consume more feed and, therefore, grow faster. Further, they convert this feed more efficiently than any chicken of a generation ago. Improved breeding is indeed an important factor in our increased efficiency of poultry production.

2. *Better Nutrition* — Research pointed the way toward more efficient feed formulations containing high energy and just the right balance of amino acids together with other essential nutrients. Where natural sources of protein did not contain this balance, synthetic methionine, one of the essential amino acids, was used to advantage in perfecting rations.

3. *Confinement Rearing* — Instead of raising farm flocks as we did years ago, we now have specialized growers with thousands of birds highly confined in houses with a considerable measure of climate control through insulation and forced-air ventilation. Modern systems of rearing birds reduce the incidence of disease and keep costs at a minimum.

4. *Better Disease Control* — This is probably the most important single factor in our improved efficiency of poultry production. Diseases were always severe, even in the old farm flock days, but with confinement rearing they would be horrendous if we did not have our present vaccinations and other medications. There is a strong interaction between good disease control and the other factors that have made for improved poultry production. Producers could hardly

afford the extra cost of hybrid birds without effective control of diseases, and certainly there would be great losses if improved feeds were wasted on birds which became moribund, died before slaughter or were rejected by meat inspectors. Confinement rearing would be utterly impossible if it were not for the improved systems of assuring poultry health developed in recent years.

Coccidiosis is a universal disease of poultry and probably the one that would put the poultry raiser out of business the fastest if suddenly today's protective medicaments were not available. With coccidiosis, birds become lethargic, do not gain weight, and death losses are often high. The chemical compounds that research has given us for the prevention of this intestinal infection are truly remarkable in their effectiveness at very low dosages when mixed with the feed. Because of the tendency of the protozoic organisms causing coccidiosis to become tolerant to the several available chemicals employed as medicaments, they are often used in rotation for a period of a few months with the objective of minimizing the intensification of resistance.

# 16

## The Case of the Wormy Grain

How long since you have found worms or weevils in flour or in a cereal product in your kitchen? Chances are you're like one housewife who replied when aksed this question, "I've never heard of such a thing." But your grandmother knew all about it. Our advanced food sanitation and modern packaging techniques, together with fumigation, have all but eliminated insect infestation of the groceries you buy at the supermarket, but in earlier times, raisins that "walked" and cereals that "crawled" were commonplace.

Unfortunately, much of the less-developed world has only begun to utilize the commodity protection technology that is available. It has been estimated by authorities on the scene that more than 15 percent of all the rice and other cereal crops raised in the Orient are destroyed by rats or insects, either in the field or in storage. Fifteen percent of Asia's cereal crops must be equated with nearly fifty million acres, an area nearly the size of the state of Kansas. Losses to rodents are potentially great everywhere, but control programs—involving sanitation, rodent-proof storages, and also the use of chemical rodenticides as one important phase—are reasonably successful where they are undertaken on a continuous basis. Grain insects, too, can be controlled through appropriate measures.

Stored grain pests have been known since ancient times.

Supplies placed in sealed tombs in ancient Egypt were found to have been destroyed by some of the same species of insects with which we are familiar when the tombs were opened centuries later. Early American colonists suffered great losses from grain insects, and the colonial army of the American Revolution is reported to have had nothing, at times, but wormy flour on which to subsist.

One of the most troublesome species, the Angoumois grain moth, became widespread during the sixteenth century as a result of global exploration and the international commerce that followed. So serious was this pest during the eighteenth century that it was the object of some of the first recorded research on insect control methods. One early French investigator described the horrors of having to rely on grain badly infested with Angoumois grain moth:

> The larvae are sometimes so numerous that if we squeeze a handful of wheat there is expressed a white, viscous fluid which is the substance of the crushed insect's bodies. The bread which is made from the infested wheat, chiefly when the flour has not been suitably bolted, contains some debris of dead bodies and excrement of the insects. It has a disagreeable, loathsome taste, sticks to the throat, lacks cohesion, and falls apart in water as if it were made of dirt. To this unhealthy food is attributed a bad throat condition which prevailed for several years, to an epidemic extent, in the countries affected by the l'alucite. This illness is characterized by gangrenous ulcerations which form in the back of the throat. The patient succumbs in a few hours, even before it has been possible to administer help.

Over fifty years ago, carbon bisulfide was found to kill insects in grain without imparting off-flavors. Research on this and the more recently discovered fumigants has shown that only inorganic residues remain after proper aeration. Through the use of sanitation along with appropriate structures and handling methods, as well as various grain-protection chemicals, losses of stored grains can be largely prevented without exceeding the levels of residues that international food safety authorities have established as permissible. Probably no type of pesticide has benefited so many people over so long a period as stored product protection materials.

Instances of fumigant injury to workers in grain storage and handling facilities have invariably resulted from disregard for recommended safety measures, just as most industrial and transportation accidents result from carelessness.

Going back to the defenseless days before modern chemicals helped protect our food is not an option that many would find nutritionally, aesthetically or economically desirable. Nor would a return to "the good old days" make sense environmentally. More cropland would have to be farmed to make up for the food and fiber that would be destroyed without modern preservation methods. In technically advanced countries, efficient protection of food in storage releases millions of acres for optional uses—forests, pastures, recreational areas, and wildlife preserves.

# 17

## Saving Energy and Conserving Soil

If you were close to the agricultural scene during the first half of the century you will recall the old farmer's dictum that one must prepare a "nice smooth seed bed" before planting. When I was a boy in the twenties it was the custom to use a turning plow, then a disk, and finally to go over the field with a spike-tooth harrow to break up as many lumps as possible before seeding.

A windstorm, followed by my usual procrastination, made me wonder if all this work was really necessary. One fall the flat roof of a small hog house blew off and landed on a grass sod nearby. The roof was old and hardly reusable, so I was to chop it up for kindling as a part of my regular chores of keeping the wood box filled in winter. There must have been plenty of kindling from other sources; in any event, it was the next June before I got around to cutting it up. The spot of grass sod it covered was completely dead, but I thought nothing of this, as it was in an unused area next to the barn. I was supposed to use a scythe on the area once or twice during the summer, and when I got around to that job in July, late as usual, some annual weeds were growing vigorously. Must we not plow and har-row repeatedly to make plants grow? How come these weeds—lamb's quarters and ragweed, as I recall—were growing so well? I didn't cut them, just to see how well they would do. Rains were abundant that summer and, by fall, these weeds were just about the tallest and most vigorous ragweed and lamb's quarter plants I had ever seen.

Why all the fuss about tillage if plants grew so well in a dead sod that hadn't been touched by a plow since the barn was built at least forty years earlier? Of course you can't cover a whole field with hog house roofs or, for that matter, with anything else that would kill existing vegetation on large areas. And the roof treatment took time. Sod doesn't die quickly when light is excluded. In those days we had no practical way to kill sod or weeds before planting except by plowing or some other form of tillage.

During the three decades following my hog house roof observation, much progress was made by crop production research people in reducing the amount of tillage needed for major farm crops. William Faulkner, in his book *Plowman's Folly* (Grosset and Dunlop, 1943), raised many questions in the minds of agriculturists regarding the need for so much tillage and the wisdom of farming slopes highly subject to erosion. The U.S. Soil Conservation Service, started in 1934, did much to alert farmers to the risks in some of their tillage practices, particularly on sloping land. The dust bowl days of the 1930s encouraged the wider use of stubble mulch methods in drier wheat-growing regions, a scheme of killing weeds present at planting time or in a fallow year by running blade-like sweeps through the soil and leaving crop stubble and weed remains on the surface where they would help prevent wind erosion. Dr. Ray Cook and his colleagues at Michigan State University developed what he called minimum tillage methods (especially for corn), a scheme that used the moleboard plow followed by a disk and sometimes a planter on the same rig. Thus, trips over the field and resulting soil compaction—which inhibits the best root growth—were minimized.

I was aware of these developments but had all but forgotten the hog house roof episode when another experience again made me wonder. In the early 1950s while experimenting with dalapon, then a new herbicide effective on grasses, I noted that, where adequate doses were applied to an old sod, smartweed and pigweed soon came up and grew vigorously all summer. Examination of their root system suggested that a dead sod was ideal for root development. Why plow and insist on a smooth seed bed if you can kill existing vegetation with herbicides, and why cultivate between rows if you can control subsequent weeds with the new selective weed-control chemicals that were then coming into use? Couldn't one raise crops without tilling the soil at all?

Early the next spring, my colleagues and I decided to find out. We killed existing vegetation (grasses and broad-leaved weeds) with dalapon and 2,4-D and then, in May, seeded plots of corn and soybeans. Weeds that came up between the rows were controlled with selective herbicides, and that fall we harvested some excellent crops, fully as good as those growing in adjacent plots that had received the usual soil preparation and inter-row cultivation. That October we put out trials with winter wheat, with equally promising results the following summer. Visitors had a hard time believing that this productive land hadn't had a plow pass through it in more than twenty years.

As new herbicides became available, crop researchers in many areas began to experiment with the growing of crops without tillage. Now, after a quarter century, over eight million acres, mostly corn and soybeans, are grown in the United States by no-till methods. Less tractor fuel is used and—even though herbicides require energy for their manufacture—there is sometimes a net savings in energy requirements. In their excellent review article in *Scientific American*,* no-till researchers Triplett and VanDoren of the Ohio Agricultural Experiment Station estimate the amount of fuel required to establish the crop "can

*Grover B. Triplett, Jr., and David M. VanDoren, "Agriculture Without Tillage," *Scientific American* 236 (January 1977): 28-33.

be reduced by as much as two-thirds of what would be consumed in conventional tillage." Research in Kentucky* indicates that further energy savings can be realized with no-till methods for corn on well-drained soils through the more efficient utilization of nitrogen fertilizer, the manufacture of which requires the greatest single energy input for that crop.

In the mid and lower-South of the United States, with their long growing seasons, soybeans can now be planted in wheat stubble the same day the grain is harvested and then weeds killed by a spray following the planter. This method not only assures fuel economy and increases total productivity of our land but also helps improve soil by keeping a crop growing on it throughout the year. Crop residues are major factors in building fertility, and double-cropping can have soil-improving benefits as long as the necessary mineral nutrients are supplied.

Much of the no-till acreage is on hilly land subject to soil erosion, and these methods are significantly reducing the silting of lakes and streams. As with any new technique, there are problems which need further research. Some corn insects such as cutworms are discouraged by plowing, and if populations get out of hand in no-till fields they must be controlled by an approved insecticide. Much research is under way around the world on no-till cropping, and situations where it offers a positive benefit in energy conservation and in preserving precious soil should soon be well defined. In some instances it appears likely that no-till crops should be rotated with those receiving more conventional treatment. No-till methods show great promise for the tropics as a means of avoiding the soil degradation that so often occurs when certain tropical soils are exposed to wind, rain, and sunlight.

Reduced tillage, whether involving no-till methods or conventional soil preparation with less cultivation made possible by the use of herbicides, will make an increasing contribution to

*R. E. Phillips et al., "No-Tillage Agriculture," *Science* 208 (June 6, 1980): 1108-13.

energy conservation. When combined with other practices that assure good crop nutrition and protection against pests so as to obtain high yields, these farming methods—undreamed of until a few decades ago—will give us the greatest return on our land and energy investment.

In considering the agronomic and economic benefits of selective herbicides it must be kept in mind that no amount of inter-row cultivation can control those competitive weeds growing in close proximity to the crop. In a ten-year series of tests with a corn-soybean cropping system at the University of Illinois, yields averaged 23 percent greater where herbicides plus one cultivation were employed, as compared with three cultivations and no herbicides.

In considering the environmental aspects of selective herbicides, one must remember that it takes almost as much energy to produce a poor crop badly infested with weeds as it does to produce a highly productive one. It must also be borne in mind that poor yields contribute to the extension of crop production on hilly land subject to excessive erosion. Results: increased silting of streams and lakes and less acreage available for permanent pastures, forests, recreational areas, and wildlife preserves.

# 18

## What Crop Is That?

My introduction to the vast spring grain belt of the United States and Canada came as a student one June in the 1930s while on a field trip with a college agronomy class. Every field was yellow with wild mustard. This weed was so thick you could only guess whether it was wheat, oats or barley until you actually went into the field and inspected the grain plants at close range. As we got out of the Model A Fords transporting us on the trip, we would make bets as to what crop lay hidden beneath the canopy of competitive weeds. The way small grains are grown, there is no chance for cultivation, and in the vast acreages of the prairies hand-pulling was unthinkable even in those days of a good farm labor supply. Ten, twenty, and even forty percent of the potential yield often gave way, year after year, to mustard, ragweed, pigweed, lamb's quarters, and other weeds.

Today there are over a half billion acres of small grains in the world, and even this does not always produce enough, judging by occasional shortages. If we went back to the weed-infested grain of the 1930s we wouldn't be just a little bit short; there would often be a food crisis of considerable magnitude. More than 200 million acres of the world's highest-yielding grain receives a selective spray of 2,4-D, MCPA or other herbicides. These astonishing chemical tools, that make weeds fade away or stunt them beyond recovery without hurting the crop, have increased yields of sprayed acreage by 10 to 20 percent and, on some fields, even more. My estimate is that bread grains that are treated with selective herbicides yield at least five bushels more per acre, making a net increase of no less than a billion bushels annually. Many authorities would say that this is too conservative a figure. That is almost as much wheat as is on hand as a reserve in the world's graineries at the beginning of some harvest seasons. A bushel of wheat will produce enough flour for at least 65 one-pound loaves of bread, so this gain in food supply through the use of selective herbicides gives us the equivalent of *65 billion* loaves, nearly 15 loaves for each person on earth.

Let's look at selective weed control in small grains in the United States and Canada. Last year, tens of millions of acres were sprayed with herbicides with the benefit of higher yields, more rapid drying and thereby more efficient harvest with fewer losses. The cost of the treatment to the farmer was generally equivalent to less than a bushel of grain per acre, so the return on his added investment was very good indeed. In fact, many fields would hardly yield enough to cover expenses if it were not for the extra bushels provided by selective weed control.

The original selective grain herbicides—2,4-D and MCPA—are still important, but recently our arsenal of weed control compounds has been expanded by the addition of several new chemicals that provide wider spectrum control not only of the broadleaf weeds but grassy weeds as well—for example, wild oats. It now appears certain that, in the future, weed control in grains can be even more complete than in the past, and productivity compared with the old days will be even further enhanced.

Bread grains are basic to our health and to our economy. Much of the world's land suitable for wheat and other cereals is already being cropped. Even thought there are sometimes surpluses we need to increase yields per acre even further, through the use of every available bit of energy-efficient technology, in order to assure future abundance.

Insects, fungus diseases, viruses, and other pests all take their toll of wheat, barley, rye, and oats. Although many have been conquered through the development of resistant varieties, modified cultural practices or biological methods, pesticides are still sometimes needed as part of a management program for these cereal crop pests. When it comes to weeds, however, we have very few choices, other than the use of selective herbicides, for closely spaced crops that cannot be mechanically tilled.

Traditional preplant tillage or working soil in a fallow year is now recognized as having the objectionable features of encouraging wind or water erosion, and of contributing to soil compaction. Stubble mulch farming, in which crop residues and dead weeds are left on the surface for soil protection, is now widely used in the drier grain-growing areas. Recent research indicates that, to avoid energy-demanding tillage even further, soil may be protected by the use of appropriate herbicides prior to seeding or before the grain has emerged.

# 19

## Plenty of Nitrogen –
## Except Where You Need It!

That was the introduction to the subject of nitrogen nutrition of plants given by my college soil science professor.

Now, this opening exclamation would be less appropriate—at least in the technically-advanced parts of the world, and today's teacher would certainly emphasize the progress of recent decades in assuring good nitrogen nutrition of crops. Little did my old prof of the 1930s realize what phenomenal progress industry would make in the economic production of nitrogen fertilizer nor what research would contribute in the 1970s to our ability to keep it in the root zone to prevent its becoming lost to drainage water or the atmosphere.

As he pointed out, it is ironic that so much of the air we breath is nitrogen gas but that ordinary plants cannot use it in this elemental form. It must first be "fixed" by reaction with some other element, with hydrogen, for example, to form $NH_3$—which we call ammonia. Since nitrogen is an essential building block for protein synthesis and proteins are vital to the formation and functioning of every cell, plants must absorb large amounts of nitrogen compounds to produce abundantly. Animals obtain their protein directly or indirectly from some form of plant life.

Since the dawn of agriculture, ten thousand years ago, a

shortage of available nitrogen in the soil usually has been a limiting factor in crop production. Rotations that include legumes which support nitrogen-fixing bacteria on their roots, the return of nitrogen-containing organic materials such as manure, and sometimes the use of supplemental nitrates from natural deposits were our only means of supplying the necessary nitrogen to the soil before today's fertilizers became available. But there was seldom enough for balanced plant nutrition and what we today consider good yields.

Then man learned to do what some microorganisms could do all along; he found a practical way to combine atmospheric nitrogen with hydrogen to form ammonia. This compound, directly or as a derivative, became an important fertilizer substance. Our remarkable increases in the yield of corn and other crops are due, to a considerable measure, to better nitrogen nutrition through the use of ammonia or compounds derived from it.

Although ammonia factories now go a long way toward eliminating nitrogen as a limiting factor in crop production, the hydrogen they use in the manufacturing process comes from natural gas. I calculate that the nitrogen fertilizer used annually in the United States is equivalent energywise to a tankful of gasoline per capita. This is a small part of our total energy, but obviously we must learn to do with as little as we can, consistent with good agricultural production, and to use the nitrogen fertilizer we manufacture as efficiently as possible. No doubt manure and other organic supplements will be more widely used in the future. Hopefully further progress will be made in encouraging natural fixation so as to reduce demand for nitrogen fertilizer requiring energy for its manufacture.

But nitrogen fertilizer will still be needed. In a tight energy situation, we will have to give priority to agricultural needs if we are going to eat, and now is the time to learn how to use fertilizer with maximum efficiency. It is estimated that at least one-third of the world's food supply is possible because of the availability of nitrogen fertilizer.

Many steps have been taken toward minimizing nitrogen losses—for example, split applications by the farmer and slow release forms such as you may have used on your lawn. Now nitrification inhibitors, a new class of agricultural chemicals, are available to help prevent the leaching that sometimes takes nitrogen fertilizer into ground water below the root zone and to minimize the conversion of fertilizer to volatile forms that may be lost to the atmosphere.

Before seeing how a chemical can accomplish this important act of conservation, a quick review of the nitrogen cycle may be in order. Let's begin with decomposing organic matter in the soil, including plant remains, manure, and other organic supplements. After several decomposition steps, the ammonium ion is formed, but bacteria soon transform this, first to nitrite, and then quickly to nitrate. Once the nitrogen in the ammonium or nitrate form is taken up by plants, it becomes a building block for the proteins so essential to growth. When organic matter is returned to the soil, its proteinaceous fraction in turn degrades, to resume the cycle.

Now, what has all this got to do with a pesticide? Until a few years ago, the answer would have been, nothing. But now crop producers have a specific bacteriostat available which—when applied in mixture with anhydrous ammonia or with a dry or liquid fertilizer—will greatly retard the growth of the nitrifying bacteria responsible for the ammonium-to-nitrite step in this phenomenon. These organisms are an important part of natural systems, and every healthy soil contains them, but in the small zone of earth where fertilizer is injected they can be a "pest," because they rapidly convert the ammonium form of nitrogen fertilizer to nitrite, which is then quickly transformed to the highly soluble and leachable nitrate. In most soils, ammonium ions do not leach following heavy rain, while nitrates do.

Nitrapyrin,* the inhibitor most widely employed (often called a nitrogen stabilizer), greatly retards this conversion process. Leaching into ground water is thus reduced. Crops produced with nitrapyrin-spiked fertilizer grow better following heavy rainfall, because they still have nitrogen available: it hasn't leached below the zone of root development.

It is particularly important to have enough nitrogen available to nourish a crop late in the season—for example, during seed formation. This is critical to high protein levels in grain and high yields in a range of crops from cabbage to tomatoes. There is a large mass of plant tissue at this stage and, if most of the nitrogen fertilizer we have applied washes down below the root zone in a heavy rain early in the season, both yield and nutritive quality can be reduced. Delaying nitrification of ammonium to leachable nitrate by use of nitrapyrin helps assure adequate nutrition toward the end of a crop cycle.

There is yet another benefit to crops from nitrification control. Nitrite formed in the first conversion step is toxic to plants and, in wet soils where it does not quickly change to nitrate, enough can accumulate to adversely effect growth. The yellowing of corn or beans you may have seen in low spots following a heavy rain is often a symptom of nitrite toxicity, and its occurrence can be reduced by fertilizing with ammonia stabilized by a nitrification inhibitor. Also, there are certain root diseases that do not thrive when the nitrogen is in the ammonium form, and they can be kept at a minimum by using ammonia that has a nitrification inhibitor incorporated with it before application. This new type of agricultural chemical will have wide scale benefits to food production in future years.

Reducing the likelihood of leaching, in addition to its economic advantages, has a definite environmental benefit. Nitro-

*D. N. Huber, et al., "Nitrification Inhibitors—New Tools for Food Production," *BioScience* 27 (1977): 523-29.

gen compounds are essential for the growth of aquatic plants, and some of the weedy conditions in lakes and streams can result from loss of fertilizer through the soil. There have been instances of nitrate levels in well-water being too high for safe drinking, sometimes the result of leaching from fields to which fertilizer nitrogen was applied. So, this new type of pesticide that stabilizes ammonia by reducing bacterial action in the zone of soil where fertilizer is applied can help slow down eutrophication of lakes and streams and minimize the chances of causing drinking water to be contaminated.

There is yet another environmental reason why spiking ammonia fertilizer with a nitrification inhibitor to slow its transformation can benefit us all. Remember that in the nitrogen cycle the first step in the conversion is to nitrite. In well-aerated soils this compound is short-lived, and a further change from nitrite to nitrate takes place quickly. But when soils are wet, this nitrite can "denitrify" and liberate elemental nitrogen and also nitrous oxide into the atmosphere. There goes valuable fertilizer that required energy for its manufacture.

The generation of nitrous oxide through denitrification, particularly when soils are wet, is a perfectly natural process, but if we increase the amount of nitrogen in the soil by fertilizing with ammonia that has not been stabilized with a nitrification inhibitor, we can be contributing to atmospheric pollution whenever a prolonged wet spell occurs before the crop has adsorbed all the fertilizer. No one is really sure what effect nitrous oxide has on the ozone layer, but some authorities believe the potential for harm is there. Thus, slowing down the nitrification of ammonia with an inhibitor can have an environmental benefit.

Like all new pesticides, nitrapyrin was thoroughly researched for safety before commercial introduction. The compound remains intact, following placement in soil, long enough to do its job, but degrades by the end of a growing season. Toxicological investigations and residue determinations using highly refined analytical techniques revealed no hazardous residues of the parent compound or its breakdown products in edible plant tissues. In case you may be concerned about the environmental effects of inhibiting nitrifying bacteria with nitrapyrin, you need have no fears, because only bands of soil are treated. The bacteria in these bands eventually recover from the inhibitory effect of the chemical. Remember that normal flora continues to inhabit the untreated portion of the soil and then reinvades the treated bands—but not until after the crop has developed sufficiently to absorb the ammonium ions from the applied fertilizer.

# 20

## Twelve People – One Grave

We once lived in an old stone house in eastern Pennsylvania across the road from a prerevolutionary graveyard. One of the aged headstones marked a mass grave for twelve people who died of yellow fever after fleeing Philadelphia, thirty miles away. Early American histories tell of outbreaks of this dread disease in East Coast ports, with the terror-stricken population fleeing to the country. Sailing ships brought both the ill and the mosquito that transmitted the fever into the ports but, of course, no one at that time knew of the role of insects in disease transmission. Little did they realize that they might already be infected when they fled the city.

Other mosquito-transmitted diseases come much closer to home than the tropical yellow fever. Malaria was once very commonplace, even in northern areas. Worldwide, malaria was the number one cause of death until a few years ago when DDT and other insecticides, combined with drainage of breeding sites, brought the transmitting mosquito under control. Unfortunately, in several countries malaria is again on the increase because of laxity in mosquito control measures.

Local outbreaks of encephalitis in the United States in recent years are a reminder of the insidious nature of insect-transmitted diseases which are so very common in the tropics. Plague, river blindness, dengue fever, schistosomiasis, Rocky Mountain spotted fever, and many other diseases can best be prevented through the control of pest vectors or intermediate hosts such as rodents or snails.

Drainage of mosquito breeding sites, sanitation to minimize rodent populations, and other environmental control practices are of utmost importance if we are to protect our health. Chemicals have also played a significant role; for example, DDT for mosquito control. Now, several other insecticides are available for use where DDT is rejected because of resistance or, as in the United States, where it has been outlawed because of possible environmental hazard. Where programs have proven effective and are already established, nothing but sheer misery could result for the populace if pesticides suddenly became unavailable. Who wants to go back to the agonies of malaria, the horrors of typhus or the finality of yellow fever? Who wants the fear that would grip us if we had no insecticides or rodenticides when a case of plague was identified?

Insecticides have often proven useful in preventing human disease by direct application to individuals infested with lice or ticks. DDT prevented an outbreak of typhus in Southern Europe during World War II by killing fleas on civilians as well as military personnel. Thus an epidemic similar to the one that killed three million people in Russia and Eastern Europe from 1918 to 1922 was prevented.

The classical text *Tropical Medicine** presents data on no less than fifty diseases of man that are transmitted by insects or other arthropods. Here are some of the more important:

*Malaria*: occurs in a broad tropical and subtropical belt around the world. It is transmitted from infected to healthy people through the bite of any one of several species of the *Anopheles* mosquito.

*Hunter, et al., *Tropical Medicine*, 5th ed. (Philadelphia: W. B. Saunders Co., 1976).

*Yellow Fever*: found in the tropics and subtropics of Africa and the Americas. Like malaria, this disease is transmitted through the bite of mosquitoes, primarily species of the genus *Aedes*.

*Encephalitis*: another important disease directly inoculated into the blood stream when a mosquito bites a healthy individual after earlier having bitten an ill one. Related forms of encephalitis affect horses and birds. Members of the mosquito genera *Culex* and *Aedes* have been identified as vectors.

*Dengue Fever and Hemorrhagic Fever*: transmitted through the bite of certain species of the mosquito genus *Aedes*. Like the preceding three on this list, they are exclusivly insect-borne.

*Filariasis* (also known as *Elephantiasis*): another exclusively-mosquito-transmitted disease of man, found around the globe in the tropics and subtropics. Members of four different genera are believed to be involved. A related mosquito-transmitted filarial species is the heartworm of dogs, common in the United States.

*African Sleeping Sickness*: caused by the trypanosomas organism and transmitted by the tsetse fly. A related disease of domestic animals which limits their production in central Africa is also carried by this fly.

*Spotted Fever*: a tick-transmitted disease found in North America. It is directly inoculated into the bloodstream by the bite of this insect, which is its exclusive vector. Similar tick-borne fevers are found in the Mediterranean region, South Africa, and Latin America.

*Leishmaniasis*: a widespread disease of tropical countries. It is caused by a tiny parasite transmitted from one person to another by the bite of a sandfly.

*Chagas' Disease*: caused by a trypanosome organism and

transmitted by the Kissing Bug; occurs in South and Central America.

*Typhus* (of the endemic and epidemic types which attack rodents as well as man): largely transmitted by the rat flea. The causative *Rickettsia* organism is deposited by the flea in its feces and scratched into skin or wounds from the insect's bite.

*Relapsing Fever*: transmitted by ticks and lice; affects both man and rodents. It is found in Africa, the Mediterranean region and parts of the Americas.

*Plague* (of the bubonic and pneumanic forms—known as the Black Death in medieval times): transmitted largely by the oriental rat flea. Rodents serve as a reservoir of infection and their fleas transfer to humans when the rodent dies.

*River Blindness* (a black-fly-transmitted disease found in central Africa and some Latin American countries): the cause of an increase in acute eye infections in recent years. As the fly breeds only in flowing water it occurs primarily in settlements along rivers.

*Scrub Typhus* (also known as *Tsutsugamumshi Disease*): a mite-borne infection found in Asia and the South Pacific. Field mice are a host to the mites and serve as a reservoir of infection. The incidence of scrub typhus was high among military personnel in the South Pacific during World War II.

Biting insects are not the only pests that can jeopardize our health. Several infectious diseases—including typhoid fever—can be carried from one place to another by flies and roaches. There are also bacterial infections that contaminate food damaged by rodents, while others can be transmitted by certain birds. Schistosomiasis, the intermediate host of which is a tropical aquatic snail, belongs on any list of major pest-related human diseases. Sanitation and other precautionary measures are useful in controlling all of these, but pesticides are often needed if human health is to be adequately guarded.

# 21

## The Devil's Lapdog

The two cats living in our barn were supposed to keep rats and mice in check, but they were not always up to their job. Dad often caught rats in traps, but in spite of this two-phased effort, we would occasionally see rats scampering away as we entered the barn, their droppings being a grim reminder that—even with vigilance—eradication of these food-destroying and disease-carrying pests is next to impossible. Once, when Dad and the cats became complacent, eggs in the henhouse and even baby chicks disappeared and the tight wooden bin in which chicken feed was stored suddenly had holes gnawed in the lower edge. The cats were encouraged to do more hunting by a reduction in their milk ration, and Dad took the next step in integrated control by supplementing his trapping with poison placed in strategic runways. Our rat problems subsided, but I am sure they were not entirely eliminated. Even if the campaign had been 100 percent successful, there would soon have been a reinvasion from a neighbor's barn.

My real introduction to the terrible food destruction by rats was in the tropics. Like others in temperate zones with interest in agriculture, I had read about the devastation caused by tropical rats that destroy crops in the field and even fruit on the trees. I had seen statistics indicating that, in many years, a third of all the food produced in tropical areas is lost to rodents in the field or in storage. But it is utterly impossible for one from the North to comprehend the rat problem in the tropics until he has spent time in agricultural areas and villages, as well as the cities. I was intrigued on my first visit to the International Rice Research Institute in the Philippines by the low electric fences surrounding each experimental plot to keep rats out. These barriers make it possible to harvest all the rice and thus obtain accurate yield data in the drive toward genetically superior varieties and better cultural methods. I understood why they would go to this trouble, having once had a part of a season's research nullified by bird depredations of my own experimental wheat plots. Naturally, a researcher must get accurate data uninfluenced by losses to pests.

Then, for a few days I was taken to farms and villages in the Philippines and really had my eyes opened. Some rice farmers had been entirely put out of business by field damage alone. Rats started consuming the new crop even before it was possible to harvest it. Many growers lost half their crop and, except for those with especially tight storages, there were further losses after harvest. Only by diligent efforts—clubbing, trapping, and poisoning with rodenticides—can many growers in the tropics produce enough to survive. The terrific losses in India and other South Asian countries exceed the food shortfall in their poorest crop years. Thus, the grain that has been shipped as gifts or under concessionary sales arrangements no more than makes up for rodent losses. No wonder this worldwide enemy of man has been called the Devil's Lapdog.

Urban rodent populations in northern latitudes can be kept down by sanitation: storing all garbage in tight cans and doing everything possible to deprive rats of food and shelter. Such environmental control is the backbone of any successful management program. Trapping and the use of rodenticides are only the final step—although important ones. In rural areas, however, depriving rats of food becomes more difficult. How do you keep rats away from feed troughs?

In the tropics, where rats attack crops in the field, how do you keep their population down, except by vigilant extermination programs? Even in urban areas in the tropics and subtropics, how do you deprive rodents of food when fruits and nuts are available in people's yards throughout the year and edible wild plants grow in every waste space? There is no answer other than a continuous integrated program using every available tool. The more difficult it is to deprive rats of their food, the more important are rodenticides as a part of an integrated program. Regardless of the part of the world involved, there is no substitute for mechanical exclusion for the protection of grain and other foods. But even with "rat proof" metal or masonry buildings, multipronged programs to keep populations in check are important.

Rodents have their parasites and predators, one of a surprising nature. Several years ago I was involved in planning environmental safety tests for an experimental herbicide which, among other potential uses, was promising for grass control in rubber and oil-palm plantations in South Asia. Imagine the chagrin of my toxicologist colleagues when I informed them that the Malaysian government recommended safety tests with cobra. Apparently this snake keeps rat populations in check, and any chemical used for insect or vegetation control must not jeopardize its numbers. Rats can do great damage to the palm nuts from which oil is obtained. My colleagues were happy when we arranged to have the tests run in Kuala Lumpur.

Early rodenticides were toxic to man and other mammals, and great care was needed to place baits where only rodents would have access. In the United States, for example, fluoroacetamide—known as Compound 1081—may now be used only in sewers and only by certified commercial applicators. In recent years the safer anticoagulant-type of rodent control chemicals have come into prominence. Mixed with grain or some other food readily eaten by the target species, these materials act by inducing internal bleeding. As the toxic bait must be consumed for several days for a lethal effect, there is less likelihood of injuring a pet or child or some desirable form of wildlife that might have a single encounter with it.

The discovery and development of the rodenticide known as warfarin forms one of the most interesting chapters in the story of agricultural research. This anticoagulant chemical, which causes fatal internal bleeding, has been a major weapon against rats and mice since its discovery at the University of Wisconsin in the 1940s. Investigating why cattle often hemorrhaged and died after eating spoiled hay, scientists isolated and identified the chemical dicoumarol as being responsible. Warfarin is the synthetic relative of this compound, widely used during the past two decades for rodent control, and now second-generation anticoagulant rodenticides, effective against warfarin-resistant species, are becoming available. Millions are indebted to this research—and to the subsequent medical developments—for the anticoagulant drugs so useful in reducing arterial blockage which have come from it. The recent emergence of rat populations resistant to warfarin emphasizes the need for continuing research aimed at the discovery of other toxins, as well as alternate methods of control.

The importance of improved methods and tools for rat control for the future can best be emphasized by a summary, written by Thomas Y. Canby in his article on rats in *National Geographic*,* of the many ways this enemy affects us:

> This year in the United States alone, rats will bite thousands of humans, inflicting disease, despair, terror. They will destroy perhaps a billion dollars' worth of property, excluding innumerable "fires of undetermined origin" they will cause by gnawing insulation from electrical wiring.
>
> In a world haunted by threat of famine, they will destroy approximately a fifth of all food crops planted. In

*Thomas Y. Canby, "The Rat, Lapdog of the Devil," *National Geographic* 152, no. 1 (1977): 60-87.

India their depredations will deprive a hungry people of enough grain to fill a freight train stretching more than 3,000 miles.

Around the world rats and their abundant parasites will spread at least twenty kinds of disease, from typhus to trichinosis to deadly Lassa fever. No less than ten types of human infections may be transmitted by rodents including fungi, cestodes, bacteria, nematodes, protozoa viruses, spirochetes, trematodes, rickettsial organisms and assorted ectoparasites. In Asia, Africa, and the Americas—including the United States—people will die of plague, the dread Black Death that destroyed no less than a quarter of the population of medieval Europe.

In several tropical nations rat populations will suddenly explode, and rodent hordes will devastate the land. Recently they overran vast areas of the Philippines, Venezuela, and the African Sahel, ravaging crops, chewing up irrigation pipes, even girdling trees in reforestation projects.

The rat problem and even the mouse problem will never be completely solved. Hopefully we will have new tools that will make population control more successful, but without diligence, month in and month out, unacceptable food losses and disease transmission will prevail.

Improved rodenticides can only come with sustained research, year in and year out, and they are most likely to originate in commercial laboratories where new compounds are synthesized and tested by the thousands for pharmaceutical or pest-control applications. In recent years picayunish regulatory requirements for premarketing approval have discouraged such organizations from even evaluating their compounds as possible rodenticides. Thus the incentive for invention provided by government-sponsored patent systems has often been nullified through over-regulation by other government agencies.

# 22

## No Bugs in Your Catsup

Have you ever tried to raise cabbage or broccoli or some other vegetable of the Brassica group without insect control? Well, I have, and more times than not the crop was ruined.

Summer and cabbage worms go together. There is no way to escape the small butterflies that lay eggs on the leaves of most members of this important group of vegetables. Soon the green worms hatch. If you have just a few plants, with patience they can be picked off. But it takes keener eyesight and more nimble fingers than the old method of controlling potato bugs merely by knocking them off the vines into a bucket with kerosene in the bottom. Cabbage, cauliflower, and broccoli would disappear from market shelves throughout much of the year if insecticides or man-applied biological control agents were not available.

In the old days, these crops were dusted with ashes or lime, both very inefficient insecticides which only slightly discourage the worms. When I was a boy we applied a dust containing calcium arsenate. It worked, but now several pounds per acre of an arsenical does not seem desirable compared with biological control involving a spray of one of the products containing *Bacillus thuringiensis*, or a few ounces per acre of one of today's specific insecticides. With proper timing of application, all are effective on the worms that attack this group of plants, and all were well checked out for human and environmental safety before governmental approval was granted.

Great effort is being made to breed vegetable varieties resistant to various diseases, insects, and nematodes. Progress is being made, but as of today pest control agents are still often needed for acceptable yields and to assure produce free enough from damage that it will be accepted by you and hold up well in storage, shipment, and store display. Federal pure food laws require that processed fruits and vegetables be relatively free of fungus and insect fragments. You can be sure there are no appreciable fruit fly wings or legs in your tomato catsup, worm fragments in your canned corn or beetle eggs in your frozen asparagus.

Without crop protection chemicals to supplement other methods of control, the inevitably lower average yield of marketable produce of many species would result in higher prices, and some items could not be produced clean enough to pass food inspection. Because most insect or fungus-damaged fruits and vegetables do not ship or store well, price increases would be particularly severe during seasons of the year when local fresh produce was unavailable. Like fruits, health-giving vegetables that add so much zest to our meals would often be out of reach of our already overstrained food budgets.

When earworms attack field corn, some kernels will be destroyed while others never have a chance to develop, because pollination cannot take place after the silks are devoured. But, if worm populations are modest, the percentage of damaged ears will not be great and the slight reduction in the yield of dry grain will hardly be missed. Sweet corn, however, is a different matter. Worm populations often seem higher in this vegetable crop than in its field corn cousin. A partially filled ear, or one with worm damage at the tip, is not very acceptable to the consumer. Then, worms keep on feeding down the ear after harvest, so several inches may have to be discarded. Fortunately, in the North the earworm is not so prevalent and appears mostly late in the season. But in the South, growers have found no way to control the pest other than by several insecticide applications between the time of first silking and harvest.

Like corn in the tropics, sweet corn raised in Florida for the fall, winter, and early spring market is particularly vulnerable to injury by earworms and the larvae of other Lepidopterous species. The following table indicates the kind of control obtained with two different insecticides in a test at Belle Glade. There were five applications at three-to-four-day intervals in this crop grown for fall harvest:

| Insecticide | Pounds active ingredient per acre* | Percent injury-free ears |
|---|---|---|
| A | .45 | 99 |
| B | .50 | 98 |
| Untreated check | — | 46 |

*1 pound/acre = approximately 1 kilo/hectare.

Note that, in the untreated plots, less than half the ears were free of injury. Under these conditions, the cost of sorting—together with the loss in marketable yield—would have made a crop grown without protection entirely unprofitable. Earworm control measures are essential to the farmer, if he is to stay in business, and to you, if quality sweet corn grown in areas where this pest is prevalent is to be available in your market. Corn borer, too, sometimes causes significant damage to ears and must be controlled if a product suitable for marketing is to be obtained.

Recently I counted up the kinds of insect pests that, at one time or another, attacked vegetables in my home garden in Michigan; it totaled sixteen. Of course there are occasionally others, but this list, I am sure, would cover most northern situations. As one goes south the number tends to increase. Revised Pest Management Guide No. 12, "Control of Insects,

Diseases and Weeds in the Home Vegetable Garden," published by the Extension Division of the Virginia Polytechnic Institute, lists twenty-two different kinds of insects common in food gardens in that state. Those who raise fruit at home would often have another half dozen. If one likes a good lawn and adds chinch bug, white grubs, and sod webworm to the list to be controlled, plus rose chafers, canker worm, and others that are primarily pests of flowers, trees, and shrubs, they would have a list of 30 or more insect species that may be encountered on home grounds.

Then there are slugs and snails, mites and nematodes, and plant diseases caused by fungi, bacteria, and viruses. Add rabbits, mice, deer, and birds and you have quite a range of damage symptoms to become acquainted with. Of course some pests are less likely than others to be present in damaging numbers, while many are a problem in only occasional years and on specific plants. Fortunately, not all localities have all kinds of pests, but even in the best of years a gardener has to defend himself against at least a few species or accept some measure of loss.

If you choose to garden without pesticides, I suggest you obtain information from local sources or learn by trial and error which garden plants usually escape serious damage in your area. Then look into available biological controls such as *Bacillus thuringiensis* for cabbage worms and ladybird beetles to keep aphids in check. Your county agricultural extension agent is a good source of information for local conditions. You may find it of interest to read again Chapters 3, 4, and 5. Gardening can be fun without taking pests too seriously if you accept some losses in good spirit.

But I think it is more fun if you can defend your plants when the need arises. If slugs are bad in a rainy summer, get some bait. When blackspot begins to show up on your roses, a fungicide approved for this pest will help. If you live where bean beetles are prevalent, a spray or dust of a recommended insecticide when the larval stage first shows up can make the difference between a good crop of beans and a miserable flop. In chinch bug country, a properly timed insecticide treatment by a homeowner or custom applicator, in years when the pest is bad, can spell the difference between an acceptable lawn and one where the grass is so damaged that weeds take over.

Yes, gardening can be fun with or without pesticides. I will not try to sell you on chemical control of pests in your home or garden, but I hope you will not try to promote nonchemical methods for protecting commercial crops against pests for which there is at present no effective alternate control. To hold out false hopes for efficiently managing pests without integrating pesticides with nonchemical methods is a sure way to contribute to food shortages in the future.

# 23

## Cockroach Country

I once heard a pesticide antagonist say that cockroach control with chemicals wasn't necessary, that with good sanitation there wouldn't be many of these pests. Obviously she lived in the North where roaches are not abundant, and from her address, I deduced that she occupied an apartment several floors up. Anyone who lives in a warmer climate would laugh at the notion that sanitation alone will keep cockroaches effectively in check. The South is Cockroach Country. Some species live outdoors; shine a flashlight on a tree trunk at night and you will see them scurrying away, in response to their light-hating instincts. But like flies, they can get in your house whenever a door is opened.

What would be the reaction of this antipesticide consumer advocate if a roach joined her at a restaurant table? You would have them in the soup, too, if restaurateurs did not diligently supplement sanitation with a residual roach-controlling spray in every nook and cranny around the perimeter of their kitchen and food storage area.

Using a low-volatile material as a coarse spray so no fine droplets float about, commercial pest control people serving food handling establishments—and homes, too—keep a roach-toxic level of a safe pesticide in all those dark places where this pest loves to hide.

When I hear the expression "the good old days," I recall the roaches in Alabama where we lived years ago when I was on the faculty of Auburn University. In spite of diligence and the careful use of the mammalian-toxic roach insecticides—which were the only kind then available—we had at least a few roaches in our home all the time. They ate the glue from book bindings in my office, got in glassware at my research laboratory, and grudgingly scurried for cover whenever I went into the adjoining photographic darkroom.

My wife, fresh from the North, felt a little less guilty about an occasional roach in our house after she saw a big one amble across the ceiling at the lovely home of the university president where she had been invited for tea one afternoon. I am sure my wife—and the president's, too—will join me in preferring the "good new days," when an occasional perimeter spray of a modern, safe insecticide approved for roach control will keep them in check—in fact, almost eliminate them as a household pest. We should know, for again, during part of the year, we live in the South, where on tree trunks just outside our door one can see large roaches at night with aid of a flashlight. A few are certain to get inside, but an occasional dead roach is all we ever see in our home. Those dark corners and under-cupboard spaces they prefer as a way of escaping the light all have a residual spray every few weeks. Aside from the suspected disease-transmitting potential of roaches, life is just a little more pleasant if one doesn't have to share one's home with them.

Among the many nuisance insects, some, like flies and roaches, are believed to transmit food-poison organisms. Others, like noninfectious mosquitoes, just drive you nuts. Anyone who has encountered a nest of fire ants, with subsequent agony from multiple bites, will hardly agree that this pest is "no more than a mere nuisance," as sometimes claimed.

Poison ivy along a pathway, an insect called silver fish in the bookcase, a wasps' nest under the eaves, mice in the pantry, and a stream of tiny ants coming in the window over the sink—I suppose human life would go on if we only controlled them the best we could without an assist from chemicals. We can do a lot by exclusion with screens, by sanitation, digging, trapping, and swatting, but appropriate chemicals for killing or repelling these irritating pests make life around our habitations a little more pleasant. Some wisely applied control programs, such as the use of mosquito larvacides, reduce nuisance insect populations at the source and thus benefit the entire community.

# 24

## The Miracle of Corn

Botanists still have many uncertainties as to the ancestral origin of corn, known as maize throughout much of the world. The plant probably arose in the highlands of South America, Central America or southern Mexico. Through thousands of years of cultivation and selection, man and nature working together performed a miracle. No other plant combines adaptability to so many conditions and utilizes the sun's energy for photosynthesis so efficiently. Corn formed the food base of the Mayan civilization and many other early American peoples.

Our modern civilization is no less dependent on this phenomenal crop, and if its origin was a miracle, so is the "package" of corn technology that has been put together during the last half-century. In spite of its utility through thousands of years and its important place in the U.S. economy, yields were modest by present standards and never seemed to change until recent decades. Note the 100-year history of corn yields per acre in the United States in the following graph. Because of drought and extremely hot weather in some areas, 1980 yields were lower than for the record year of 1979, but still not far below the average for the 1970s.

How did this modern corn miracle come about? Why, after the better part of a century with average yields of less than thirty bushels per acre, were we able to increase them so re-

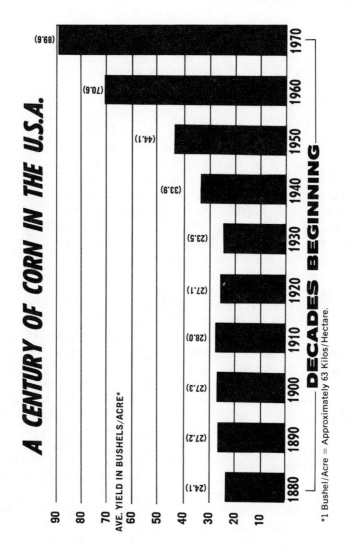

A CENTURY OF CORN IN THE U.S.A.

AVE. YIELD IN BUSHELS/ACRE*

90
80
70
60
50
40
30
20
10

DECADES BEGINNING

| 1880 | 1890 | 1900 | 1910 | 1920 | 1930 | 1940 | 1950 | 1960 | 1970 |
|------|------|------|------|------|------|------|------|------|------|
| (24.1) | (27.2) | (27.3) | (28.0) | (27.1) | (23.5) | (33.9) | (44.1) | (70.6) | (89.6) |

*1 Bushel/Acre = Approximately 63 Kilos/Hectare.

markably? Let's look at the anatomy of this phenomenal increase in the productivity of corn, a crop that, in addition to syrup, cooking oil, cereal products, and alcohol, indirectly provides us with dairy products, eggs, poultry, pork, and grain-finished beef.

It is sometimes assumed that our modern farm equipment, which minimizes labor inputs, is the major factor in corn yield increases. Mechanization is indeed responsible, to some extent, because it permits the planting of large acreages at the optimum time and also the harvesting of crops when conditions are just right. But with adequate labor and favorable soil moisture, one could still obtain very high yields without modern equipment, providing they satisfied the other requirements: 1) highly productive hybrids, 2) good nutrition through fertilization, and 3) efficient control of pests. All of these are important. Let's look at them individually.

It is estimated that modern corn hybrids at a high level of nutrition and with good pest control have the genetic capacity to out-yield old corns by at least 25 percent. They not only produce more tonnage of grain per acre but also have strong roots and stalks so the plant stands up well for mechanical harvesting, an important contribution to high yields in our mechanized agriculture.

Unless already present in the soil in adequate amounts, the major nutrients—and sometimes minor ones, too—must be supplied as fertilizer for the outstanding corn productivity we have today. Conservation farming—involving rotations including legume crops and the return of animal excrement to the land—enabled better growers of older days to produce fifty to seventy bushels per acre with the varieties then available. Now, with superior hybrids and more adequate fertilization to supplement good conservation farming practices, the same farms are producing 125 to 175 bushels per acre, and occasionally one tops the 200-bushel mark. Actually, today's farmers are building fertility of their soil with these huge crops, not only through the addition of essential nutrients but also through the large residues of organic matter returned to the soil as roots and stalks. In a sense, they are growing manure in place.

But the most productive hybrid with the best fertilization can fall short of optimum yield if it is not protected from pests. The universal use of fungicidal seed protectants to permit early planting and, thereby, the growing of long-season hybrids has already been mentioned in Chapter 9.

Many kinds of insects attack corn in one locality or another, and insecticides are often necessary to make investments in hybrid seed and fertilizer pay off. In the tropics leaf-eating worms, stem borers, and aphids often attack when the plants are small: the bugs seem to be there just waiting for the seedlings to emerge. In temperate zones, insects are not as severe, but when varietal resistance, rotations, and cultural practices will not provide adequate control, insecticides can be very profitable in terms of dollars and in efficient use of land and energy resources. Cutworms often damage corn stands soon after emergence, and an insecticide must be used to avoid severe losses. Corn borer, too, sometimes reaches numbers that could cause unacceptable yield reductions without an application of an appropriate insecticide.

One of the more widespread insect pests in the United States is the corn rootworm, the larvae of a small beetle. Roots severely damaged keep plants from full development and make them more subject to lodging. No matter how good the ears, corn harvesting equipment cannot gather them if stalks are bent over out of the path of the harvesting machine. Rotations help keep rootworm populations down, but sometimes adequate rotation is not feasible, and an insecticide is needed.

Look again at the yields of corn before we began to use insecticides for better rootworm control, say before 1950. A part of those increases in yield—of the decades of the 1950s and 1960s, and our even more productive period of the 1970s— reflects better root systems supporting healthier, more erect plants, the result of the use of effective soil insecticides. Ten bushels per acre greater yield on 15 million treated acres is the most conservative estimate I have seen on the value of rootworm control. This equals 150 million bushels in the United States.

Actually, significant acreage expansion over corn plantings of recent years would tend to be on sloping land ill-suited to row crops. Erosion would increase with resulting silting of streams. Thus, elimination of corn rootworm control with insecticides, in our present farming system, would have indirect adverse ecological consequences. Until we perfect a better method of control, insecticides banded on the soil over the row at planting time appear to be the best answer where rootworms are present in numbers adequate to cause an appreciable yield loss. Aerially-applied insecticides to kill the adult beetles prior to egg-laying as a method of disrupting their life cycle is another method that shows promise in some areas.

Although itself a grass, corn is tolerant of several herbicides that control many grassy weeds as well as the broadleaved types. Today's high yields could not be achieved with appreciable competition from weeds, and the only alternative to the present wide-scale use of selective herbicides would involve a return to inter-row tillage, with its tendency to prune crop roots, compact soil, and encourage erosion. But there would still be competitive weeds close to the corn plants not handled by the cultivator. Hand hoeing, to be effective and safe to the crop, must be done within a short period. If my recollection of the amount of corn I hoed in a day as a youngster is correct, I calculate it would take at least ten million people three to four weeks in the spring to hoe the seventy million acre U.S. crop. When I contemplate ten million temporary farm workers, I recall the difficulty apple growers have had in recent years in rounding up a few thousand pickers for autumn harvest— which in my view is a much pleasanter job than hoeing corn.

Obviously there is no practical way to produce today's

high corn yields without selective herbicides. Hopefully the several now approved from which growers can choose the herbicide most effective for their particular weed population will be supplemented, in the future, by still others that combine the highest level of selectivity, weed control efficacy, and environmental safety.

# 25

## More Contented Cows

Farming has always had its ironies: too much rain at hay-making time and, weeks later, dry weather as corn tassels—just when the crop is in greatest need of moisture. Another irony struck me when I was a boy in a farming community in southern Michigan. When spring grass was lush and dairy cows should have been giving the most milk of any time of the year, the heel fly showed up. Sometimes cows spent more time and energy chasing around the pasture trying to escape the fly than they did eating grass and clover. Milk production often went down, in spite of abundant forage. Those flies were laying eggs on the hide of the lower part of the animal's legs, obviously a disturbing experience. Cows must truly be contented to be highly productive, and the heel fly precluded that state of bliss.

Cattle also suffered from large internal grubs that made lumps on the animal's back during winter and reduced carcass quality at slaughter. The holes in the animals' backs, made by the grubs as they worked their way out, resulted in a lowered hide value. Who would want grub holes in their shoes?

Entomologists finally put together the many acts of the heel fly and cattle grub drama. Now we know that these two insects are but different life stages of one and the same pest. After spring fly activity, tiny worms hatch from the eggs deposited on the hide of the lower legs and then gradually migrate between the muscles to the region of the gullet, where they reside for some time. In the fall and early winter these larva make their way, again internally, to the animal's back, where growth is completed and breathing holes are made in the hide. Come late winter the grubs—often called bots or warbles—drop out of the hole and fall to the ground where they spend the pupal or resting stage of their life cycle. When weather warms and spring grass is lush, the adult flies emerge from the pupal cases in the soil. Thus the life cycle is completed and the cattle chase is on again.

But livestock entomologists began to dream. They visualized the possibilities of an insecticide so safe to the animal that it could be injected or fed after the heel fly had deposited her eggs and the grubs had hatched. If the parasite could be killed and resorbed within the animal's body at an early stage without hurting the bovine host, the debilitating effects of the later grub stages could be avoided, but more important, the cycle could be broken and there would be fewer heel flies the next year. Fortunately not many other animals harbor this pest; if all the bovine animals in an area could be treated with this hoped-for systemic insecticide, the heel fly and grub population would be very low indeed.

For decades, livestock entomologists sought such a systemic "grubicide" without success. Then some of the second generation organo-phosphorous insecticides proved effective against the larval form of the insect without causing harm to cattle. One of these remarkable insecticides could be given orally in feed at very low doses over a period of several weeks, while another formulation was effective as a spray. Then researchers had a happy surprise when it was discovered that, with proper formulation, some of these materials could be applied from a dipper to the animal's back and still achieve good control. After exhaustive tests for possible residues in milk and determination of a safe slaughter time following treatment, to assure meat that would present no human hazard, these new "pour-on" grubicides came into wide use.

So successful were these materials, the pest was actually eradicated in Ireland. Elsewhere, the pest is being kept at low population levels by use of approved systemic insecticides in years when heel fly activity in the spring forecasts a high population of internal grubs, and now more contented cows are producing more efficiently than ever before. In Latin America the dreaded Nuche or Torsolo—an even more debilitating species related to the nothern cattle grub—is now being controlled by these same pesticides. Thus, milk and meat are made more plentiful in many areas where protein is badly needed by an often malnourished population.

# 26

## A Cornucopia of Parasites

Grubs are not the only insect enemy of cattle that can increase the cost of producing beef and dairy products. Animals covered with flies of one kind or another fail to gain weight or produce milk efficiently. To be highly productive, a bovine animal must spend all its waking hours eating and ruminating. Anything that distracts it, including worrisome flies or other pests, will result in lower feed or forage intake. Some insects, like horn flies, ticks, and mosquitoes, actually suck blood and are therefore especially hard on growth and milk production.

The damage that can be done by horn flies was recently demonstrated by experiments in Nebraska in which a large group of brood cows, with their calves, had to pass under dust bags containing a fly-control insecticide in order to get to their watering troughs. A similar group on an equally good pasture had no fly control. When calves were weighed in the fall those from the horn fly control group averaged nearly 13 pounds more than those from the unprotected herd and, at that time, were worth $5.75 more. The insecticide cost about $1.00 per animal, thus making a net return of $4.75 per head. And there were some hidden benefits; heavier calves are sturdier and stand shipping better if they are being moved and are better able to withstand the winter if being held on the range.

The increases in weight were believed, by the researchers

doing the work, to result from better milk production by the mother cows during the summer months. It was estimated that the untreated brood cows often had more than 1,500 horn flies feeding on them, while the protected cattle seldom had more than 50.

A new approach to fly control involves a topical "spot" application with an insecticide that is safe to the animal and persistent enough to greatly reduce the fly population over several weeks as the insects move from one part of the animal's hide to another. Recent research has resulted in a promising systemic insect growth regulator for horn fly control. When administered orally, it will greatly reduce fly populations over a considerable period and thus give livestock producers another useful tool in their efforts to make production more efficient.

When the summer fly season ended, in days before good livestock insecticides were available, life for a bovine animal hardly became a bed of hay and grain. With cooler weather, cattle lice often took over as the pest-of-the-season. Weight loss and poor general condition by the end of the winter was partly due to this debilitating parasite. Sprays have been used in the past, but wetting down a critter in cold weather has some obvious risks to the health of the animal: cattle get pneumonia too. With some of the newer insecticides, back rubbers have become effective. Cattle are inclined to rub their backs on anything that is handy, especially when they are infested with lice. Insecticides dispersed by back rubber devices, placed for convenience of the animal, have proven surprisingly successful under many conditions. A "pour-on" formulation is also available; one dipperful on the animal's back provides louse control over a lengthy period. Now a new "spot-on" treatment using a minimal dosage is giving good louse control.

Various kinds of ticks are among the prime enemies of cattle in many parts of the world, particularly those species that transmit such diseases as Texas Fever. Control by sprays and dips of approved chemicals can be successful if properly done.

Some areas, including parts of Australia, would long since have ceased beef production were it not for pesticides to control disease-carrying ticks.

Cattle are not the only livestock species that suffer from pests. Hogs, poultry, and sheep all have their louse species. Mange mites similar to those that occasionally attack your dog are likewise troublesome on sheep. These animals also have woolworms, keds, and nose bots. Mites and lice infest both chickens and turkeys; literally a cornucopia of parasites. All can now be successfully controlled with approved insecticides. No dairyman or livestock or poultry producer who lets the bugs get far ahead of him is likely to be in business long.

The screwworm, which can attack various kinds of livestock, is being eliminated in some areas by the release of sterile male flies (Chapter 4). But a spray for treating worm-infested wounds is still essential in places where this biological control program has not been carried out or is not 100 percent effective.

If you traveled in Florida years ago, you may have shared my impression that cattle were ill-adapted to that state. Seldom did one see a sleek, well-fed animal, even when the forage was lush. Today, Florida is a large producer of cattle, and most have the appearance of being healthy and well nourished. Intestinal worm control had a lot to do with the change. The tropics and subtropics can raise good cattle only when periodic medication for parasite control is practiced. Hogs and sheep in all climates have their internal worm parasites, and any producer who neglects administration of one of the specific control chemicals, called anthelmintics, is likely to suffer severe losses through slower weight gains and less efficient feed conversion.

Flies around farmyards, poultry houses, feedlots, and milking areas have long been of concern to public health officials. Sanitation, including manure removal, is of utmost importance, of course, but sprays, baits or the new insect strips for farm

use are very helpful. In early commercial development is a new
type of insecticide of the hormone type that can be fed in
extremely low doses and will prevent the flies—which go
through their larval stage in manure—from ever becoming
adults.

Your pet—whether dog, cat or horse—is also fair game for
a multitude of different internal and external parasites that are
now kept within bounds by pesticides and veterinary drugs.
The old expression, "a dog's life," hardly applies in these days
of flea collars, systemic control of mange mites, medication for
heart worm, and effective insecticides for ticks.

I consider that flea collar on your dog quite an invention.
The insecticide used to impregnate the plastic collar must have
just the right physical properties to disperse in the plastic resin
and remain there for some time, yet be held loosely enough
that the flea will pick up a trace when it wanders to the region
of the neck. If you recall pre-flea collar days, when we obtained temporary control by bathing the dog using a special
insecticidal soap, you will agree that the chemists who worked
this one out made a unique and valuable discovery. Environmentally, the flea collar is important too; we no longer flush
insecticides in the bath water down the drain. The pill you give
your dog to prevent heart worms during the mosquito season
is another useful product of research on parasite control.

There are at least ten million horses in the United States and
significant numbers in other prosperous countries where this
former beast of burden is now a companion animal. Many of
the same insect species that attack cattle also make a horse's life
miserable during certain seasons. In addition, this animal has its
own special pests. Approved insecticides for fly control around
horse premises make life more pleasant for both the animals
and their caretakers. Some insect control agents are recommended for direct application.

Why be concerned with animal parasites and diseases? Some
would argue that livestock and pets should be protected on

humanitarian grounds. Agriculturists are likely to point out how necessary it is for the farmer to prevent losses if he is to survive economically. Others may stress that livestock and poultry products account for a sizable part of our total food bill and anything that can keep them from becoming more expensive will have anti-inflationary as well as nutritional benefits. Some economists might argue that, in the interest of balance of payments, an agricultural country should avoid losses in every practical way and thus keep food imports at a minimum.

All of these are valid arguments, to which I will add conservation of energy and land resources. Remember that livestock and poultry require many pounds of grain or forage to produce a pound of edible product. Conversely, a modest improvement in efficiency through the avoidance of losses is magnified manyfold, in terms of crops saved and the cropland required to feed us. Every acre farmed requires the expenditure of energy. Protecting livestock against parasites and diseases is an indisputable conservation measure. Reduced demand for feed and forage means that land best left in forest or desired for recreation or wildlife preserves is less likely, out of necessity, to be diverted to agriculture.

# 27

## Ancient Food Crop – New Technology

Rice is a remarkable food crop both agronomically and nutritionally. It can be raised on ordinary soil; each year, millions of acres are grown like wheat or oats in Brazil and a number of other countries. But unlike these temperate-zone cereals, rice will thrive when the land is flooded. This swamp-plant trait has permitted its culture on river plains where heavy soil and chances of flooding restrict the kind of food plants that can be grown. Purposeful flooding, where water is available, helps control weeds and assures adequate moisture at all times. The flood water is drained off, at the end of the growing season, to permit harvest.

Rice from flooded "paddies" has been a mainstay of the diet of people throughout much of the tropics and subtropics for thousands of years. Even countries as far north as Spain, Italy, Japan, and South Korea produce large tonnages of rice, and this cereal is an important part of their diets. Most people in the Orient and many in Africa and Latin America depend on locally-grown rice for the bulk of their food supply. Today about 60 percent of the world's people look to rice each day as a major source of calories. Rice is especially important in Asia, where three-fourths of the world's undernourished live.

Rice is well adapted to storage and transportation, and only dehulling is required to make it ready to cook. When supplemented with protein from animal products or legumes, together with fruits and vegetables as a source of minerals and vitamins, rice makes a highly satisfactory basic food. Many northerners prefer bread and potatoes as a prime source of energy, but those who grow up on a rice diet prefer it to other starchy foods.

Much rice in less-developed countries is still grown as it has been for millennia—without benefit of fertilizer or pest control measures. Yields are similar to those obtained in the past; with favorable weather, around 1,500 kilograms per hectare (approximately 1,500 pounds per acre). But rice in the Mediterranean basin, the United States, Japan, Taiwan, South Korea, and parts of Latin America yields three or more times this amount per hectare. High rice yields have contributed significantly to the food abundance of these areas. Now, an increasing number of farmers throughout all of Asia are utilizing improved high-yielding varieties, together with the plant nutrition and crop protection technology that must accompany these genetic improvements if significantly enhanced yields are to be realized.

Better rice varieties adapted to local conditions and tolerant of local pests are now being developed in many parts of the world including Asia and Africa. Alone, these genetic improvements contribute to local food supply and when, in addition, the economy permits the use of mineral fertilizers to fit a given soil's deficiencies, there can be another incremental increase in yield.

Pest control chemicals can hardly pay off in traditional rice culture, but as crop vigor improves with fertilization the toll of pests in terms of weight of grain lost per unit area increases. A farmer with a productive variety that has been fertilized for high yield cannot afford to allow insects such as the brown plant hopper to keep his crop from realizing its full potential. Even where this insect does little damage by direct feeding on rice plants it can have a disasterous indirect effect by transmitting virus diseases. Those who have invested in the several inputs essential for a high level of productivity can ill afford to allow rice blast or other leaf diseases to devastate a crop. The ultimate goal is the development of varieties highly resistant to the major pests of a given area, but until this feat is accomplished, the use of crop protection chemicals will sometimes be essential.

Put it this way: a 20 to 40 percent increase in a low-yielding rice crop may not pay for good pest control, while a similar percentage increase in a crop with a yield potential of 4,000 to 5,000 kilograms per hectare can be profitable to the grower and of value to his country's economy, as well as help assure enough food for her people. Conversely, a 20 to 40 percent loss of a crop, in which investment for seed and fertilizer has already been made, may spell financial disaster.

Look at the data in the following table from a test at the International Rice Research Institute in the Philippines. The last three varieties listed in the left hand column have been bred for insect resistance.

| | Grain yield in kilograms per hectare | |
| Variety | Without treatment | With insecticide treatment |
| --- | --- | --- |
| Taichung Native | 550 | 3,470 |
| IR8 | 1,300 | 4,920 |
| IR22 | 3,630 | 5,560 |
| E583-1 | 4,740 | 5,840 |
| E597-3 | 5,220 | 7,440 |

This experiment demonstrates the remarkable progress that is being made in breeding varieties for resistance to insect damage. At the same time, it illustrates the fact that, for top yields, an insecticide may still be needed. Under the conditions of this experiment, only the integration of genetic and chemical methods allowed the rice to realize its full potential. The more productive numbered varieties in this test are not yet commercial ones, but are being evaluated further and made available to local breeders throughout the rice-producing world. Obviously, the rice in this experiment was well fertilized and weeds were adequately controlled.

Fungus diseases have long taken a significant toll in rice production in many areas of the world, in spite of considerable resistance on the part of older, local varieties and even more resistance in modern types resulting from efforts of the plant breeder. Now, recently-developed chemicals differing considerably from the older fungicides are helping growers obtain the full measure of grain they should realize from their efforts and various inputs. Unique compounds produced by cultured microorganisms and classified as antibiotics are now widely employed as a part of integrated programs for the control of certain leaf diseases. In the future, growers will have systemic seed treatment chemicals available for prevention of seedling infections (Chapter 9).

Upgrading the productivity of rice in less-developed areas must be a gradual process involving improved genetics, better crop nutrition, and appropriate insect, disease, and weed control measures. Often improved storage, transportation, and distribution must parallel these production improvements, if the rice farmer is to be assured of a market. Without the incentive of a market for his grain, and at a price which he has reason to hope will net him a profit, a rice farmer in the technically less-developed parts of the world has little reason to gamble on modern inputs, regardless of how badly food might be needed.

In the technically advanced countries, where most rice is direct-seeded, rather than hand-transplanted as is still practiced in Asia, selective herbicides are very essential. Hand-weeding has become out of the question where there is not an abundance of agricultural labor. Without weed control, yields would be entirely inadequate to pay production costs, let alone give the grower a chance of a profit. If herbicides that control grassy and broadleaved weeds without hurting the rice were not available, there would be little of this crop grown in many countries where it is now an important food, i.e., France, Italy, Spain, Colombia, Venezuela, Brazil, and the United States. Herbicides are increasingly important in several countries still using some traditional cultural methods—including Japan, Taiwan, Korea, and the Philippines.

Ever since the term Green Revolution was applied to the progress less-agriculturally-advanced countries are making toward greater food crop productivity through the use of improved varieties, better crop nutrition, and superior pest control, there have been critics who claim that the old rice-growing methods are best. It is true that some mistakes have been made, such as—temporarily—too great a reliance on one variety inadequately evaluated under local conditions and too slow a development of the infra-structure of storages and transportation facilities to keep pace with increased yields. It is true that small farmers have not always benefited from improved technology as much as the large ones (often because of inadequate credit). But these problems are recognized, and much effort is being expended to solve them.

Meanwhile, increased yields have contributed significantly to food supplies where they were often short. Mankind is far past the point of no return with respect to the utilization of modern crop production technology. With over 4.5 billion to feed today and prospects of at least a doubling of population, particularly in countries highly dependent on rice, there is only one alternative to vastly increased yield: more hunger and eventual starvation. Good pest control is an essential part of any effort to step up rice yields, and today chemicals are still an essential component of effective integrated pest management programs.

# 28

## Time to Emerge from the Gathering Era

Driving through the Manistee National Forest in western Michigan, one has glimpses here and there of spindly pines in straight rows planted by the Civilian Conservation Corps in the 1930s. A few of these trees in open spots have attained a respectable size for their age of about forty years, but most have been kept stunted by an overstory of scrub oak which seldom makes useful timber in that sandy soil. These retarded plantings are testimony to the folly of just setting out trees and then letting them take care of themselves.

One wouldn't expect much of a harvest if tomato plants were just stuck in the soil of a poorly prepared garden and then ignored. Competing plants that were there first or ones that grew faster at early stages (those we call weed-trees, because they are growing where we don't want them) have dominated these plantings made by the CCC, just as weeds would usually dominate an untended garden. After a long struggle, some of these pines will grow into full sunlight and eventually make respectable trees, but it will require two to three times as long as it should.

During the hunting-and-gathering stage of man's existence, the world provided sustenance for a few million people; then the development of agriculture prompted a marked growth in population, as more and more food became available. But forestry is just now emerging from the gathering era. So vast were the world's woodlands in relation to need that the supply of trees seemed inexhaustible, and natural regeneration, no matter how slow, was assumed to be adequate to take care of the future. Now, however, with man's burgeoning population and increased demand for wood and paper and energy, it is time that he emerged from the gathering era. Silviculture—the production of trees as a crop—must expand tremendously if future generations are to have the blessings of abundant forest products.

Back in the Manistee Forest, there are a few pine plantings that have been "released" from competition and are already being harvested for posts and pulpwood. After this thinning operation, the remaining trees will grow on to saw-log size. High labor requirements have kept more weed-tree control from being carried out. Worthy plans for providing work in forests for unemployed city youth have been put forth, but the logistics, costs, and organizational problems involved have prevented meaningful accomplishments.

During the last three decades, some useful herbicides for the killing of undesirable species before tree seedlings are planted have become available. Other products for selective weed-tree control in struggling stands have come into use. Where wisely employed by competent foresters, these chemicals have shown remarkable value. Labor is still required for their application, but a man with a back-pack sprayer or a tree injector containing an appropriate herbicide can be many times as productive as one carrying a brush hook, axe or saw. As in managing pastureland covered with brush, just cutting unwanted woody growth in forest situations is like going to the barbershop: it grows right back from sprouts on the roots or crown. Systemic herbicides, on the other hand, move downward in the plant to inhibit further development of subsurface buds.

"Why not do it by hand to give people needed work?" is the logical question often asked in connection with weed-tree and brush control in forests. Recent comparisons of hand versus chemical methods in the Pacific Northwest of the United States indicated that one experienced person could do the necessary work on about twenty acres in a three-month cutting season. With upward of five million acres in need of treatment in the United States each year, we are talking about a quarter of a million people to do the job. Assuming they could be employed, trained, and transported, and then housed and fed in the forest areas where the work was to be done, there are still reasons why herbicides are the better answer. Here are the views of Dr. M. Newton, Professor of Forest Ecology at Oregon State University:

> (1) cutting of brush without herbicide treatment leads to vigorous sprouting and a need for retreatment within a year or two; (2) total exposure of suppressed conifers by cutting leads to sun scald and mortality or decreased growth or released trees if more than a few feet tall; (3) slash interferes with wildlife and worker access; (4) the high accident rate suggests that this approach may have serious implications regarding human health and safety; (5) the labor requirements make accomplishing the very large task at hand possibly unfeasible with the trained labor force available; and the risk of using untrained personnel is intolerable; and (6) the cost is 2 to 25 times greater than that of a selective herbicide application.

Many essential jobs in forestry cannot be done with chemicals, and the need is there for more labor whenever increased public or private funds can be devoted to assuring wood and wood products for future generations. For example, there is a great need for increased production of forest tree seedlings, and for planting them, in appropriately prepared sites. It is my opinion that available man- and womanpower should be devoted to these and other labor-intensive phases of tree production, not squandered on weed-tree and brush control.

Chemicals can do a better job of it, at a fraction of the cost.

In many situations, ground application of herbicides is impractical and aerial spraying is the most effective method of site preparation. Application with specially designed and equipped aircraft, usually a helicopter, may also sometimes be the only practical method of selective control of weed plants that keep desirable young trees from living up to their genetic growth capacity. In several forest areas, herbicides have proven useful for keeping unwanted combustible vegetation from overrunning fire breaks and forest roads. They have also been used to desiccate tree trimmings and other forest "fuel" for safe controlled burning at a time when adjacent areas are too green to permit the escape of fire.

In addition to the use of herbicides to help with the establishment of tree plantations, foresters occasionally employ other chemicals to protect their crop. Tree seed is difficult and expensive to collect, and fungicidal protectants are sometimes applied to it before planting to reduce the risk of decay. Nursery soils abound in weed seeds and other pests, and preplant fumigation is widely practiced to assure sturdier and lower-cost seedlings. Fungi sometimes threaten the closely planted stock in tree nurseries, and fungicides can protect against devitalized seedlings, ones that, transplanted into a forest site, would have a reduced chance of survival. Rodents and other vertebrates can raise havoc with new plantings, and—if the population is high—controls must be instituted.

Insects, too, can take a heavy toll in nursery beds, and appropriate insecticide treatments must sometimes be used. Established trees being grown for timber or pulpwood can survive an appreciable insect population without undue loss, but occasionally, the threat is intolerable. Much progress is being made in the use of insect pathogens for keeping populations of certain pests within acceptable levels, but now and then an insecticidal spray must be applied by air, to assure tree survival.

Interior plywood of Douglas Fir lines the walls of the room

where I am typing these words, and hidden from view is sheathing and framing lumber of the same kind of wood. On my desk I have a small vial of Douglas Fir seed given to me by a nursery manager in Oregon as a reminder of the importance of this species to the future supply of forest products. What would it take to transform these seeds to 150-foot trees two feet or more in diameter and ready to be cut for timber and plywood? Given good care, trees from these seeds could be ready to harvest in forty to fifty years. Perhaps their sister seeds, millions of which are planted each year, will produce lumber for my great-grandchildren's homes.

But what if trees from this seed are not given the care needed to assure efficient growth? If it takes them 100 years to reach saw-log size, they will not furnish lumber until my great-, great-, great-grandchildren are seeking housing! What about the needs of intervening generations? With poor silvicultural practices, each crop of trees will of necessity occupy productive forest sites much longer than our growing population can afford. Such land, like that used for food crops, is finite; they stopped making it a long time ago.

What must be done to efficiently transform these seeds on my desk to trees ready to harvest for pulpwood, lumber, and other forest products? Nature must provide the soil and water, and the sunshine, but what must the forester do? Just scattering seeds about is unlikely to bring forth a harvest in less time than a century—if the seeds are, indeed, fortunate enough to sprout and grow at all. Let's look at various phases in the development of a Douglas Fir plantation and see where man must fit into the equation and, particularly, where chemicals are proven silvicultural tools.

The soil in nurseries where Douglas Fir trees get their start is often infested with weed seeds, nematodes, and fungi; here, a preplant fumigation assures good seedling growth. Seed protectants are sometimes used to assure a good stand; during their one or two years in a nursery bed, Douglas Fir seedlings are often attacked by insects and fungi, both of which can be readily controlled by an appropriate pesticide. Selective herbicides have been found that greatly reduce nursery weeding costs when weed seeds have not been killed by a preplant fumigation.

Long experience has demonstrated the futility of planting out fir seedlings on land that is chocked with competing vegetation, so sites must be prepared appropriately, often through a combination of chemical and mechanical methods.* Once planted, our healthy young trees would have little chance of rapid growth if too much competing vegetation were allowed to dominate the area, so a selective spray or a weed-tree injection program is often carried out on one or two occasions.

One of the promising new biological agents may be applied for tussock moth when natural factors fail to keep populations below the threshold of economic damage. But these agents are not always adequate, and an insecticide is occasionally needed, usually of the carbamate, the organophosphate or the hormone type.

So with seed treatment, soil fumigation, seedling insect and disease control in the nursery, rodent supression, site preparation, selective weed-tree control, and sometimes an insecticide to protect established stands, our Douglas Fir crop will benefit from perhaps eight to ten pesticide applications during its lifetime of 40-50 years. Considering the value of land, taxes, and management costs related to time, the economic benefit of rapid growth—to which pesticides make a major contribution—is self-evident. But aside from obvious monetary benefits are the hidden eventual costs of inadequate housing, a shortage of paper and other forest products, together with the inevitable infla-

---

*For a more detailed presentation of steps in forest tree production in western North America, the reader is referred to: *Regenerating Oregon's Forests,* compiled and edited by Brian D. Cleary, et al., Oregon State University Extension Service (Corvallis, Oregon, 1978).

tion that results when the demand of a necessity exceeds the supply. Add these economic factors to the environmental cost of inefficient land use and it becomes evident that we must manage forests of the future as a crop, utilizing appropriate technology—including herbicides and other pest control chemicals—where the need dictates.

This is not to say that "tree farms" cannot also have recreational uses. Well-managed forests are among the best recreation areas. Good management is not only compatible with recreation, it is highly desirable for all uses beyond "wilderness experience." A wider range of wildlife will thrive on the varied environment of a managed forest (older trees interspersed with recently-cut-and-planted as well as middle-aged blocks) better than on an untouched wilderness. If the superior tree-growing sites are managed intensively for wood production, with clear recreational side-benefits, land will remain available that can be set aside specifically as wilderness areas for wildlife and human enjoyment.

PART

# III

## *Pesticide Safety*

[Concepts, Controversies, and Controls]

# 29

## What Is a Poison?

*A poison* is *too much*! That is how my biochemistry teacher defined it years ago. It is still a meaningful broad definition of the term, even though, in everyday usage, the designation *poison* is reserved for those highly toxic substances capable of causing illness or death at relatively low levels of ingestion or other forms of exposure.

First for the broad meaning of the term. Everything is poisonous if taken in excess, even water and table salt. Vitamin A is essential to human health. No one could long survive on a diet completely lacking in this complex chemical synthesized by plants and present in many foods, yet in laboratory tests with rats, Vitamin A has been found to cause birth defects and to be carcinogenic when administered in large doses in their feed. Some land is unsuited to grazing, because its excess of selenium is taken up by forage plants in amounts that will poison livestock. At the same time, traces of selenium are essential to proper growth of poultry and farm animals, as well as humans.

Plants need boron. Many crops cannot develop properly without traces of this element, and some soils, where it is lacking, must be fertilized with small amounts for good productivity. Yet at higher doses, compounds containing boron are toxic to plants. Some weed killers used on railroad beds to keep all vegetation in check derive their phytotoxic properties from this element.

Now for the everyday concept of a poison: one of those substances that can result in morbidity or mortality at relatively low levels of exposure. They may have various origins. Some examples:

> Animal: snake venoms
> Green plant: the toxic chemical in nightshade
> Fungal: mushroom toxins and aflatoxin
> Bacterial: botulinus toxin
> Mineral: arsenic, lead
> Inorganic chemical: sulfur dioxide
> Synthetic organic chemical: PCB

All have similar features: 1) very small amounts appear to be harmless, 2) larger amounts cause sickness from which recovery is possible, and 3) still larger amounts may cause death.

An important unresolved question concerns the possible effects of very small amounts that cause no detectable sickness. Some people believe that, even though there is no human epidemiological evidence, the minutest trace of certain chemicals may induce cancer long after the exposure takes place. These are the compounds that have been found to increase the incidence of tumors in laboratory animals when administered at high doses. Others postulate that the body's natural defense mechanisms protect against cell damage from low doses, even though the high levels used in laboratory tests may overwhelm the system. Because of these opposing views, we will first discuss toxicity of a noncarcinogenic nature and then take up the divergent views regarding the question of a threshold versus no-threshold for carcinogenic compounds. (There will be more on this conflict in Chapter 34.)

*How much is too much?* That depends on the substance involved. With the possible exception of carcinogens, nothing becomes a poison until the threshold of toxicity is reached. Obviously, only a massive excess of water is harmful. A spoonful of common salt taken at one time could make an adult very ill, but it would require only a fraction of a spoonful of iodine as a single dose to cause severe internal damage—even though traces of this element are essential to health. Iodized table salt has long been recommended by the medical profession as insurance against goiter, an enlargement of the thyroid gland resulting from iodine deficiency. Yes, how much is too much depends on the toxicity of the particular substance in question and the role of exposure.

At the far extreme from water is aflatoxin, a naturally-occurring substance produced by a mold that sometimes grows on stored grain or nuts. The toxicity of aflatoxin is so high that the equivalent of a tiny grain of salt could cause severe illness or even death. Yet most of us have ingested traces of this substance at one time or another, at levels which had no immediate harmful effect. And here we arrive at that knotty question of possible delayed adverse response from tiny doses, because aflatoxin has been established in laboratory tests as a potent carcinogen. In fact, this naturally-occurring substance has been shown to induce tumors in tests with animals at far lower doses than do most synthetic compounds that have been ruled to be carcinogenic, yet there is no direct evidence that the traces of aflatoxin that may have been consumed in food is responsible for cancer in humans.

*How much also depends on the species.* Ocean fish thrive in salt water that will quickly kill many fresh water species if placed in it, and as we know, ocean water can also kill people if they ingest too much. Pyrethrins from the pyrethrum plant is highly toxic to many insects, yet one can safely use it for the control of household insects, because of its very low level of toxicity to man and other warm blooded animals. In contrast, a dose of naturally occurring atropine (extract of Belladonna) that would be devastating to humans has little effect on most insects.

The miticide tricyclohexyltin hydroxide will control many

species of parasitic mites at a spray concentration of only 2 ounces per 100 gallons, yet this same spray has practically no effect on insects. The dose of 2,4-D you use on your lawn to kill dandelions does not harm the grass. Conversely, a spray of dalapon will kill or retard the growth of many grasses, while alfalfa and clover growing in the same sod will be little affected. None of these herbicides has a measurable effect on insects, fish, mammals or other wildlife at exposure levels far above those incidental to their use. The old saying, "One man's food is another man's poison," might be paraphrased to say "a substance highly toxic to one species at a given dose may be innocuous to another."

*Potential toxins vary in their mode of action.* Those that have an effect soon after an adequate oral dose or other type of exposure are classified as acutely toxic. Substances or dosages that induce a response only after repeated exposure or after a prolonged period of time are said to have chronic toxicity. Some individuals lack a critical metabolic capability to handle certain substances that are normally detoxified or metabolized and are said to have idiosyncrasies to specific drugs or even foods. Others suffer from allergies or become sensitized to specific compounds with which they may have contact. Among the irreversible toxicological responses to repeated exposure to certain chemical compounds are some forms of cancer, defective development of an embryo, and gene mutations.

Many compounds are readily metabolized, particularly by warm-blooded animals; thus, low levels of exposure over a long period have no discernible effect if they do not exceed the organism's metabolic capacity. For example, in evaluating a carbamate insecticide for safety to wildlife, it was found that a dose that was lethal to ducks if given at one time had no measurable effect when mixed with a full day's diet. Further, this level of administration was continued for a full year without response. Thus, in the complete diet, 365 times a single lethal dose had no effect on growth, fertility or general health.

Other compounds are continuously eliminated through normal excretions. Liver and kidneys are key organs of purification in the mammalian body, and the amount of any potential toxin they can handle over a given time span without damage to their tissues varies with the species and, no doubt, the individual. Margins of safety must be built into allowable intakes to provide for these variations. Common ethyl alcohol is a well-known chemical which is readily metabolized but which can cause irreparable harm to the liver if too much is consumed over too long a period.

Some potential toxins accumulate up to a certain concentration in fat, often without apparent risk to the individual. Others known for their chronic toxicity—for example, the heavy metals mercury and lead—can accumulate in various organs with disastrous effects if exposure continues at a high enough level.

*Measuring Toxicity.* Substances being evaluated for their toxicological properties are subjected to experiments with laboratory animals to determine whether special precautions are needed to handle them safely. This includes tests to determine their potential for harm if swallowed, if in contact with eye or skin, or if inhaled. Acute effects are measured in terms of quantities or concentrations needed to cause an adverse response. One frequent measurement is the single oral dosage required to kill 50 percent of the animals involved, commonly called the $LD_{50}$ and expressed in milligrams of test chemical per kilogram of body weight of the laboratory animal (mg/kg). For example, table salt has an $LD_{50}$ of approximately 3000, while numbers for aspirin range upward from 1100 in various experiments on its acute oral toxicity. Among the several tests conducted with malathion, the lowest reading was 885 mg/kg of body weight while parathion, another member of the same chemical group (which is sold only to certified applicators), has an $LD_{50}$ ranging from 25 to 30 in different investigations.

These data serve as useful guides to the degree of caution

required in handling various substances, but it must be recognized that they are measures of single oral dosage toxicity only. A compound having a high numerical $LD_{50}$ (indicating low acute oral toxicity) may, on long-term exposure in dietary feeding tests, prove to have chronic effects and to present a significant hazard if carelessly used. On the other hand, some compounds that must be handled carefully to avoid ingestion, as dictated by low $LD_{50}$ numbers, show no adverse response on long-term exposure at the low levels likely to be encountered when reasonable care is exercised. The carbamate insecticide experiment with ducks referred to earlier furnishes a good example.

Long-term tests lasting from ninety days to two years in which the substance being investigated is mixed with the entire diet of the laboratory animal are among the most important phases of toxicological investigations. Test animals are weighed weekly, while blood and urine chemistry are monitored throughout the duration of the test. At termination, a complete autopsy is conducted and the histopathology of a wide range of organs investigated. It is in part based on these tests that regulatory agencies make decisions on safe daily intake levels for humans (Chapter 31). Additional tests—for one to three generations of laboratory rats and mice—are needed as a basis for judgment on questions of carcinogenicity, mutagenicity, and reproductive physiology.

*Safe levels of ingestion must be set.* For chronic toxicity studies other than possible carcinogenic effects, research people generally employ a range of doses for administration to laboratory animals in long-term feeding studies that will give a positive response at the higher end of the dosage range but, at the same time, establish a no-adverse-effect level. The choice of doses for such tests is indicated by short-term dietary feeding experiments. The substance being evaluated is mixed with the animal's complete diet. Many compounds show no effect based on growth, reproduction or longevity when administered to rats or other laboratory animals at doses, on a body weight basis, far above any likely human exposure. Neither do they reveal any pathological changes when tissues of sacrificed animals are thoroughly studied, under the microscope, by qualified pathologists. Therefore, to establish an adverse effect level with substances of low mammalian toxicity, an exaggerated dose must often be used.

The next dosage down the scale below the first to show an adverse effect is known as the highest no-adverse-effect level. For example, if the daily administration of the compound under investigation is a series of daily doses of 1, 2, 4, and 8 milligrams per kilogram of body weight, and at 8 milligrams the animals show only weight reduction but at 4 milligrams there is no deviation from the control group, then the highest observed no-adverse-effect level is 4 milligrams per kilogram. This number is important in establishing safe intake levels, whether the chemical being studied is an environmental contaminant, a natural compound, a commercial food additive or a substance, traces of which may get into food as a consequence of its production, storage or processing.

When calculating safe intake levels for people and establishing tolerances for residues in food, regulatory agencies (the Environmental Protection Agency, and the Food and Drug Administration in the United States) provide for a wide margin of safety, usually 100 to 1. Thus, if one milligram per kilogram in the total diet of experimental animals was established as the highest no-adverse-effect level, the estimated acceptable daily intake for man would be no more than .01 milligram per kilogram of body weight. Wide margins for safety are needed because of possible differences in response between laboratory animals and humans, between individuals, and for dietary variations.

Actual amounts of residues which would be established as legal limits would vary, of course, with the importance of a given food in the diet and the number of food items that might carry traces of the same substance. Obviously, if a pesticide were permitted for use only on cabbage, the residue level could be higher than if additional residues were also expected to be found in a wide range of vegetables. If residues are present in a food common to the diets of infants, the elderly or the ill, additional safety factors may be applied.

*Now for the different views regarding carcinogenic response from low doses of chemicals.* Many experts claim that data on radiation-induced cancer strongly support the concept of a linear dosage-response curve; in other words, that there is no threshold below which cancer will not occasionally be induced. These biologists, studying the effects of radiation on various organisms, visualize even a very low level as having an adverse effect on the "reproductive machinery" of an occacional cell; now and then, they postulate, one of these modified cells initiates "wild" growth, resulting in cancer in one of its many forms. They visualize radiation-induced cancer as being an expression of a permanent and replicable change in DNA— the substance in each cell that controls the nature of its multiplication.

If very low levels of radiation do, indeed, cause an occasional cellular aberration, do mere traces of carcinogenic chemicals have a similar effect, or can the body's natural defense mechanisms detoxify minute quantities before they have an opportunity to react with the cell constituents that are responsible for the control of subsequent growth? Many biologists believe that traces of carcinogens do have an effect. They visualize an occasional cell being attacked by a molecule of the chemical involved in such a fashion that it becomes malignant. It is believed that most of these altered cells are destroyed (or repaired) by natural defense mechanisms and do not result in abnormal growth, but when one does proliferate, a cancer of some sort may result. The frequent long time-lag between exposure to a carcinogen and the sometimes observed manifestation of cancer is difficult to explain. However, human epidemiological studies on the relationship of cigarette smoke and certain industrial chemicals to the occurrence of cancer strongly suggests that, the longer the exposure and the larger the dose, the greater the number of cells that may undergo a malignant transformation and, therefore, the greater the chance that one of them might cause a pathological malignancy.

On the other hand, many scientists believe that the action of a chemical carcinogen differs significantly from that of radiation. Tissues are readily penetrated by radiation, and there is nothing to prevent interaction with DNA. It is pointed out that, in contrast, a chemical carcinogen must go through several intermediate steps of transport and perhaps metabolism before it can react with the cell's reproductive constituents. The body's defenses are visualized as including membranes that do not permit traces of chemicals from coming in contact with DNA and also enzyme systems that promote DNA repair and others that detoxify foreign molecules. These scientists point further to evidence that certain enzymes apparently have the ability to identify and destroy aberrant cells produced by chemicals that slip through other defenses. In short, they believe that organisms, including man, are capable of dealing with small quantities of carcinogens, and it is only when protective systems are overwhelmed by higher doses that they present a threat.*

*Laboratory tests for carcinogenicity must be run at relatively high dosage levels.* Low levels, comparable on a bodyweight basis to probable human intake, seldom induce tumors. Even the kind of dosage range used for establishing a proven no-effect level for other than carcinogenic effects will often be too low to give a carcinogenic response. Therefore, exaggerated doses are used. For example, the lowest dose of saccharin found

*For a further discussion of the threshold vs. no-threshold controversy, the reader is referred to: "Chemical Carcinogens: How Dangerous are Low Doses?" *Science* 202:37-41, October 6, 1978.

to increase the incidence of tumors in rats was equivalent to the daily human consumption of several hundred bottles of a diet soft drink. Decisions on appropriate dosage protocols for toxicological experiments is complicated by the fact that laboratory animals are naturally short-lived. If much time is required for response to a low dosage to show up, the animal may die of old age first.

Certain forms of bacteria are known to mutate readily in response to a number of chemicals that have also been found to be carcinogenic in laboratory animals. It is now believed by many pathologists and toxicologists that most compounds which show evidence of mutagenicity will also prove to be carcinogenic in an appropriate test. The speed and simplicity of raising bacteria or other microorganisms has prompted the development of tests for mutagenicity as a shortcut to the identification of chemicals that might cause cancer. These tests are sometimes useful, but may yield false positive or false negative results. Long-term experiments with appropriate laboratory animals are still required to establish legally acceptable data for the establishment of tolerances for any substance that may be present in food.

*The Delaney Clause in the U.S. laws regulating the purity of foods* legislates against the addition of compounds that have been determined to be carcinogenic. During the 1950s, the Congress of the United States reviewed past legislation related to food safety, with the view of making appropriate amendments. Representative James Delaney of New York recommended, among other things, that no tolerances be permitted for food additives that had been found to be carcinogenic in man or in appropriate tests with animals. Thus, in recommending against such tolerances, Representative Delaney subscribed to the concept that there is no threshold below which a carcinogen may not be responsible for an occasional cancer. A majority of the Congress subscribed to this view, and the clause in the 1958 amendments that outlaws additives with a carcinogenic potential is known as the Delaney Clause. The wording of this portion of the law, with additional discussion, will be found in Chapter 33 and also in the final part of this book, Trade-Offs. It was in compliance with the Delaney Clause that the banning of saccharin was threatened—even though the effect on laboratory rats interpreted as being carcinogenic occurred only at high doses.

In the United States, pesticide tolerances in foods are established by the Environmental Protection Agency, while animal as well as human drugs are regulated by the Food and Drug Administration. Since its passage, the Delaney Clause has served as a guide to various regulatory agencies when the establishment of permissible tolerances in foods was under consideration. Thus, in practice, it applies not only to those substances that are purposefully added to foods but also to those, such as pesticides, that are present incidentally, as a result of production, processing, storage or packaging. As the clause is currently interpreted, nothing that is found to induce cancerlike tumors in laboratory animals is permitted to be used in any manner that might result in a food residue, no matter how miniscule and no matter how high a dosage level is required to obtain a carcinogenic effect on the test species involved.

Statistical projections by some toxicologists, made from data obtained in laboratory experiments where tumors were induced at relatively high dosages, suggest the possibility that the incidence of cancer might be slightly increased in an exposed human population, sometimes by as few as one in a million or even one in ten million. Other professional toxicologists are firmly convinced that such projections are invalid and that, because of natural defense mechanisms, there is a level of intake below which there is no hazard. Neither side can prove its point; the number of test animals that would be required stagger the imagination. Obviously, we can't devote unlimited amounts of money, brainpower, and laboratory facilities to the study of a few compounds in infinite detail when there are so many other research problems related to human health crying for solution. So there is a continuing conflict of views regarding the safety of chemical compounds that have been found to increase the incidence of tumors in laboratory animals when administered at doses greatly in excess of likely human exposure. Further aspects of this controversy will be discussed in Chapter 33.

# 30

## The "Natural Is Safe" Misconception

Those whose science studies are limited sometimes fail to comprehend that all substances are made up of chemicals. The earth itself and all upon it, including the air surrounding us, is a mixture of chemicals. A few are simple elements, such as oxygen and nitrogen in the atmosphere, but most are compounds made up of different elements joined together in reactions that take place throughout the natural world. Some compounds are simple like water—$H_2O$, two atoms of hydrogen plus one of oxygen—and table salt—$NaCl$, one atom each of sodium and chlorine. Other compounds, such as cellulose—which contains hundreds of atoms of carbon combined with hydrogen and oxygen to make up the elongated molecules that form fibers of wood, paper, and cotton, as well as the basic structure of growing plants—are fantastically complex.

Many of the synthetic compounds man finds useful are quite complex—for example, aspirin, known to the chemist as acetyl salicylic acid. Others are simple molecules, like the ammonia widely used as fertilizer or the ethylene used as a "building block" to form the more complex polyethylene we recognize in plastic films and containers. Both ammonia and ethylene occur widely in nature, so man has merely learned how to shortcut natural phenomena. Many of the more complex synthetic compounds are chemically identical to ones found in nature—for example, ascorbic acid, commonly known as Vitamin C. Useful molecular structures found in nature do furnish leads which chemists can follow up with the synthesis of related compounds—for example, the pyrethroid insecticides, which are synthetic variations of the naturally occurring insect control agent, pyrethrins.

Whether synthesized in nature, a laboratory or a chemical manufacturing establishment, each compound has its own structural characteristics and its own effect on living organisms. Many furnish nourishment for one or another form of life, while others are useful as medicines or pest control agents. But most of the chemical compounds in the world around us are innocuous, for all practical purposes, because we don't eat them or breathe them, and they are not absorbed dermally. Experience or controlled experiments tell us which may have a harmful effect, and at what level of ingestion or other type of exposure.

As I look out my window while writing these words, the maples are showing their first tint of red, and soon the Michigan woods and roadsides will be aflame, all because in autumn many leaves synthesize chemical compounds of a group known as the anthocyanins. Let's look at the structure of one of the pigments of this class:

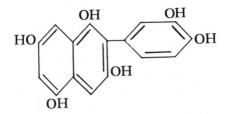

The atoms making up an anthocyanin must be fit together in a highly efficient manner by the cells of the leaves; otherwise, trees could not change from green to red so quickly following the first chilly nights of late summer and early fall. A common yellow pigment of autumn leaves, xanthophyl, is equally complex, and chlorophyl, which gives plants their green color, has an even more complicated structure. Certainly, as chemical "factories," plant cells are remarkably efficient.

For several years I followed with interest the chemical research by colleagues aimed at the production of a new herbicide, known by the generic name triclopyr, which I had had a hand in discovering. To the chemists, it is the compound 3,5,6-trichloro-2-pyridyloxy acetic acid. Many reaction details had to be worked out, first in laboratory synthesis, then pilot plant scale-up and, finally, in full production. Only recently, after years of effort at a considerable cost, is synthesis going smoothly enough to make commercialization possible. Now, with the completion of detailed studies on safety to people and the environment, triclopyr is registered and will be available for several vegetation control purposes, where it has proven safe and highly effective in extensive field tests.

As chemical compounds go, triclopyr is not a highly complex molecule in comparison, for example, with many pigments found in plants as well as their carbohydrates, fats, and proteins. Then why all this research needed on how to produce it efficiently? Compared with plants, man is a neophyte when it comes to synthesizing complex organic molecules. New techniques make it easier for the chemists to identify and map the structure of complex molecules, but in spite of many outstanding synthesis accomplishments, the most skilled chemists are pikers. A single leaf, using only carbon dioxide from the air, water and minerals from the soil, and sunlight as a source of energy, can turn out dozens of different complex organic compounds day after day. I have known chemists to struggle for weeks to produce a small laboratory sample of a single compound, and sometimes they never succeed in synthesizing the desired chemical structure.

One has only to experience a brush with nettles or poison ivy, a bout of hay fever from ragweed pollen, or stings from wasps or fire ants to be reminded that nature abounds with substances that are toxic if we have too great an exposure. The extremely potent toxins synthesized by *Bacillus botulinus* under certain conditions of food preservation and by the *Amanita* group of wild mushrooms are recognized as among the most poisonous substances known to man.

Most of the complex chemicals synthesized by plants are useful in one way or another to the plant itself; many provide the natural resistance plants have against most pests (Chapter 3). Biologically derived compounds furnish animal life with both energy and the building blocks for growth. Green plants form the first link in food chains for microbes, insects, vertebrates, and other organisms that lack the ability to utilize sunlight as a source of energy.

But not all plant-derived compounds are useful for the nutrition of man and other animals, and some are toxic, even at low doses. Long before Socrates drank the cup of hemlock, it was learned that various plants and substances extracted from them have marked pharmacological effects on people. Some, like digitalis, are medically beneficial at certain doses but toxic at higher levels, while others are of no known use to man and hazardous if ingested.

Solanine from nightshade berries, glucosides from the leaves and stems of oleander, and nicin from castor bean: these and many other toxic substances are derived from green plants. Some common house and garden species, parts of which can be hazardous if ingested, are: daffodils, bleeding heart, autumn crocus, boxwood, Christmas rose, daphne, larkspur, foxglove, oleander, Jerusalem cherry, Dieffenbachia, privet, yew, and certain lilies. Jimson weed, nightshade, and henbane are among the garden weeds that produce specific toxic chemical compounds that one must avoid ingesting. Kingsbury* lists no

*John M. Kingsbury, *Poisonous Plants of the United States and Canada* (Englewood Cliffs, N.J.: Prentice Hall, Inc., 1964).

fewer than 100 species of plants that produce specific chemical compounds known to be toxic to humans or livestock, if ingested in adequate amounts. Even though the woods and roadsides—and even our gardens—harbor these plants, they present no practical hazard, with reasonable care. Fortunately, those that synthesize dermally toxic compounds, like the 3-n-pentadecylcatechol in poison ivy and poison oak, can be avoided or controlled by appropriate methods. Plants that are poisonous to livestock if consumed in too great a quantity are referred to in Chapter 13.

Unfortunately it is a widely held misconception that naturally occurring substances—whether purified or composed of a mixture of many chemical compounds—are safe, while those that man has synthesized are hazardous. Nothing could be further from the truth. The synthetic materials that make up such everyday things as fabrics, paints, and other finishes, floor and wall coverings, paper sizings, rubber-like goods, plastic dishes, and packaging materials are no more likely to present a hazard than substances of strictly natural origin. As we have seen, many compounds derived from living organisms can be highly toxic with adequate exposure. Many "natural" substances of inorganic origin also present a potential hazard at adequate dosages. Examples are the lead compounds widely used in paints before synthetic components became available and the mercury compound once used in processing felt that led to the expression "mad hatter," because of its neurological effects.

True, certain synthetic compounds, some occurring only as impurities, are known to be carcinogenic, based on human experience or tests with laboratory animals, and are therefore under rigid controls. Some natural substances have similarly been found to be carcinogens, e.g., aflatoxin, which is also acutely toxic at relatively low levels of ingestion. Still other naturally occurring chemicals induce various kinds of irreversible responses when excessive dosages are administered to laboratory animals, e.g., the teratogenic effects of common salt and Vita-min A. Nicotine-containing insecticides derived from tobacco stems must be used with the same precautions as synthetic mate-rials sold for home garden use. Pyrethrins from the pyrethrum plant is relatively safe as a household insecticide, but so is malathion, synthesized in a chemical manufacturing facility.

With all toxic substances, natural and synthetic, dosage dic-tates the degree of hazard. Can you imagine one knowingly taking a swim in water containing a poison "laced with it," in the language of TV and newspaper reporters? This voluntary exposure would be particularly hard to imagine if it were known in advance that the substance in the water could have severe acute effects and be lethal at adequate doses, and further that some laboratory investigations have indicated it to be a carcinogen at adequate doses. Then imagine someone recogniz-ing all this, going to a restaurant after a swim and ordering food that is known to contain this same toxic substance! All this is exactly what we do when we swim in the ocean and then order a seafood dinner to top off a pleasant day, for all the oceans contain traces of arsenic and as much as 20 parts per million of this element are present in many kinds of shell fish. Yet who would disagree that a restful day at the beach with a swim in the sea followed by a delightful seafood dinner can contribute to good health?

At times we all fall prey to the temptation to make broad generalizations, based on few facts, and whole classes of chemicals have often been the victims. Parathion, an organo-phosphate compound, has a high level of mammalian toxicity and is employed for insect control only by skilled operators, but other members of this chemical group, such as malathion, are remarkably low in mammalian toxicity (see Chapter 29). Several years ago a French pharmaceutical containing the ele-ment tin in its molecular makeup was found to have unfor-tunate side effects, and the notion sprang up that tin compounds in general were toxic. The highly toxic roach poison, sodium fluosilicate, gave rise to the notion that all fluorine-containing compounds presented a hazard. An amusing aspect of both these unfounded generalizations is that a perfectly safe compound containing both tin and fluorine, stannous flluoride, is used in toothpastes for its desirable antibacterial and enamel-preserving properties.

Each molecule has its own effect on living systems, and a seemingly insignificant change in molecular structure can vastly alter biological response. The contrasting plant response to the herbicide 2,4-D as compared with its closely related isomer 2,6-D interested me years ago when I was involved in a study of the structure-activity relationships of the phenoxy group of herbicides. Let's have a look at their chemical structures:

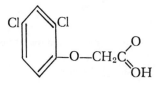
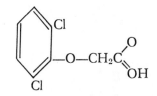

2,4-D
(2,4-dichlorophenoxyacetic
acid)

2,6-D
(2,6-dichlorophenoxyacetic
acid)

Note that the only difference is in the position of one chlorine (Cl) atom on the ring (a shift from the 4 to the 6 position), yet biologically, 2,6-D is innocuous to plants at moderate doses, while 2,4-D selectively removes weeds from your lawn and provides grain growers with a practical way to cope with yield-depressing weeds as described in Chapter 18. Obviously no generalizations about the biological effect of all compounds of the phenoxy group are justified.

In addition to positional changes, replacement of one element with another results in a new compound, often with vastly different biological properties. Examples are naturally occurring acetic acid—the sour factor in vinegar—and some of its chlorinated relatives. See what happens when one or more hydrogen atoms making up acetic acid are replaced with chlorine.

Acetic acid

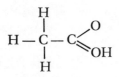

A naturally occurring food substance used for its flavoring and preservative properties and marked by low toxicity to animal life.

Chloroacetic acid

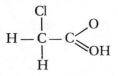

A highly toxic substance used primarily as an intermediate in chemical synthesis.

Dichloroacetic acid

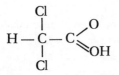

A little-used compound of low toxicity.

Trichloroacetic acid

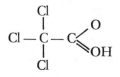

In the sodium salt form this compound is an herbicide used for grass control. It is only moderately toxic to animal life.

Note the low level of toxicity to warm-blooded animals exhibited by acetic acid and dichloroacetic acid, on the one hand, and the highly toxic properties of monochloroacetic acid on the other. Why does the substitution of one chlorine atom for a hydrogen atom so greatly increase toxicity, while a second chlorine lowers it almost to the level of acetic acid itself? Among these chemicals, only trichloroacetic acid has specific grass-killing properties. Obviously, one cannot generalize about the biological effects of this group of compounds.

Mark Twain once remarked, "All generalizations are false, including this one." In spite of his warning, I will risk the generality that *generalizations regarding the toxicological properties of groups of chemical compounds are false.* Each compound must be investigated and judged on its own merits, whether it occurs in nature or never existed until synthesized by man.

Chemicals vary in many ways other than their effect on organisms, and some of these characteristics are important in considering safety in its broadest aspects. The rate of decomposition of a chemical in the environment, the nature of its decomposition products—called metabolites or degradates—and the mobility of the compound in soil, water, and air are all properties that must be understood.

Most organic substances, those with the element carbon in their molecular make-up, are decomposed in natural environments, some very rapidly and some slowly. Factors promoting degradation are microorganisms in the soil, ultraviolet light, and the presence of water to encourage a reaction known as hydrolysis. Pest control chemicals that degrade slowly and, at the same time, can move readily in the environment must be studied carefully for their possible effect on nontarget organisms as well as on man.

Just as many reactions may be required in the synthesis of an organic molecule, various steps are involved in its degradation. And just as considerable research is needed to learn how to synthesize a compound, further research is required to understand the steps through which it breaks down to carbon dioxide and water or other substances common to the environment. Sometimes toxicological tests are required for intermediate degradation products that may be present where organisms could be exposed.

Mobility depends on many factors, including degradation rates; compounds that break down quickly have little chance to move. Water solubility, and also the way molecules are adsorbed on soil constituents, are factors. Many pesticides used for crop protection move downward through the soil very slowly and, thus, have an opportunity to degrade long before they can reach groundwater. When erosion occurs, chemicals adsorbed on soil particles can move over the surface to eventually reach streams or other waters. Some aspects of erosion prevention are discussed in Chapter 17.

Some chemical compounds tend to be soluble in water, while others with low solubility in aqueous systems are often somewhat soluble in fat. When traces of the former escape into the environment, they seldom concentrate in living organisms, but fat-soluble compounds do sometimes bioconcentrate. Because of their constant exposure to the water around them, aquatic organisms are more likely to show a higher concentration of trace chemicals from the environment than are terrestrial forms of life. Fat-soluble chemicals reach only a certain concentration level in tissues, depending on the compound, the exposure, and the characteristics of the organism involved.

There has been much speculation about the possibility of concentrations of trace chemicals found in the environment increasing from one organism to another up the food chain, a phenomenon called biomagnification. Research has now shown that this seldom occurs. The misconception that biomagnification is a common phenomenon may stem from the misinterpretation of data on concentrations found in the whole organism at one step in a food chain, but only in fatty tissues of the next organism up the line. This subject will be discussed further in Chapter 32.

I hope the highlights of chemistry and toxicology in Chapters 29 and 30 will aid in an understanding of pesticide regulation and the registration process to be discussed on the following pages. A sharpened conception of the scientific aspects of pesticide safety should also aid in an appreciation of the different sides of controversial issues and in weighing risks versus benefits.

# 31

## How Pesticides Are Regulated

Guarding against life's hazards was once purely a personal matter. Our prehistoric ancestors had to learn about dangers of the hunt from their compatriots or by experience, and they had no building inspector to help assure that their huts would not be blown down in a storm or crushed by a heavy snow. Even during my boyhood, safety decisions were largely personal. Design of horseshoes was up to the blacksmith, and when a faulty wheel came off our buggy and Mother and I were thrown into the ditch, no regulatory controls had been broken, and no inspection station bypassed.

But with the blooming of the Industrial Age, new hazards became obvious, and collective efforts at minimizing their impact were inevitable. The automobile spawned speed limits, then safety glass and seatbelts, and—more recently—exhaust emission controls to reduce pollution of the air we breathe. All other modern modes of transportation and all forms of energy generation brought forth regulations aimed at fewer accidents and the protection of public health. Structural specifications, communication systems, and the design of a wide range of manufactured goods came under regulation, all with the objective of greater human safety. New knowledge of the cause of various diseases resulted in a wide range of regulations of sanitary systems and water supplies. Meanwhile, a vast expansion in commercial food-processing brought forth regulatory controls related to impurities, additives, and the safety of containers. The occurrence of untoward side effects of certain drugs and the occasional marketing of worthless nostrums prompted legislators to establish approval schemes for all manner of remedies.

There can be no doubt that regulations have contributed much to health and safety, particularly those that promoted pure water, safe milk, and other sanitary measures. Reduced speed limits after the 1973 oil crisis were followed by significantly lower accident rates. Many factors are involved in man's increased longevity, but regulations no doubt played a part. Look at the average life expectancy in the United States:

| | |
|------|----------|
| 1900 | 48 years |
| 1925 | 55 years |
| 1950 | 68 years |
| 1975 | 72 years |

Early pesticide regulations, requiring that purveyors attach a skull-and-crossbones warning sticker to packages containing acutely toxic compounds, was mentioned in Chapter 7. With expansion in the number and chemical diversity of pesticides during the decades following World War II, many countries began to legislate or decree new regulations aimed at minimizing the likelihood of adverse effects associated with their application. The new controls were also designed to protect consumers against the presence of hazardous traces of pesticide residues that might be present in food.

In the United States, for example, the Federal Insecticide, Fungicide, and Rodenticide Act of 1947 (FIFRA), together with subsequent amendments to the laws regulating foods, required that—prior to sale in interstate commerce—pesticides be registered and a label giving clear use, directions, and precautions be approved. Preregistration efficacy tests were a requisite for approval, and various toxicological tests along the lines discussed in Chapter 29 were required to determine what precautionary statements should be included on the label.*

A major requirement of these U.S. laws, and one that is now important in registration procedures elsewhere, was the measurement of possible residues of the pesticide in question in food crops or livestock by an approved analytical method following experimental application, in accordance with the proposed label. When residues were found to occur, long-term dietary feeding studies were mandated as a basis for establishing safe levels of intake. Residue tolerances could then be established commensurate with the degree of toxicity observed and the amount actually found in the agricultural commodity after treatment according to proposed label directions.

FIFRA was amended in 1972 to include intrastate regulation and to prohibit the unauthorized use of any pesticide in a manner inconsistent with labeling instructions. Regulations since promulgated under the amended law require wide evaluation of proposed pesticidal chemicals for possible effects on nontarget species. These data, along with information about persistence in soil, water, and air, as well as mobility following application, serve as additional guides for the United States Environmental Protection Agency in determining whether a product will be registered for a given use and, if so, what restrictions may be needed to avoid undue environmental risk.

A key feature of changes brought about by the 1972 amendments to FIFRA is the current classification of the uses of pesticide products into two groups: 1) "general" uses and 2) uses that are "restricted." The former includes products considered safe enough to be used around the home and be sold

*For a history of pesticide regulation in the United States and a discussion of current regulatory procedures, the reader is referred to: "Food Safety; Where We Are," Committee of Agriculture, Nutrition, and Forestry, United States Senate (Washington: Government Printing Office, 1979).

"over the counter." Those in the restricted use category may be purchased only by farmers or commercial applicators whose qualifications have been certified, following the passing of an appropriate test.

FIFRA was again amended in 1975 and in 1978 to streamline certain administrative procedures and to clarify enforcement responsibilities by states and the Federal Government. The authority and responsibility of EPA for ordering reviews of safety of a previously registered product or use, and for suspending or canceling registrations when new data on safety dictate that such action was also clarified. The amended law provides for provisional registration when there is a current need for the product, when safety has been substantially established, and when only a few secondary tests have yet to be completed.

As amended, the law also gives states the authority to register products for use within their boundaries for special local needs, providing residue levels following such uses are within federal tolerance standards. The Food and Drug Administration has the responsibility for monitoring most foods for possible residues in excess of established tolerances and the authority to take appropriate action, including confiscation. The monitoring of red meat and poultry for pesticide residues is the responsibility of the U.S. Department of Agriculture, as part of its broader inspection program. If tolerances are exceeded, a report is made to the Food and Drug Administration for investigation and possible prosecution.

It is important to an understanding of pesticide regulation procedures in the United States to recognize that, although legal tolerances never exceed safe levels, they are often lower than that, depending on what the analytical tests reveal may be present. In other words, the tolerances that are set reflect what is likely to be present when the product in question is used according to proposed label directions, even though a higher tolerance would be safe. Only those uses which have been investigated with detailed residue determinations and for which tolrances have been established appear on registered labels. The sale of unregistered products is prohibited.

In the United States and many other countries, plant growth regulators are controlled under the same laws that govern pesticides. These remarkable chemicals have been responsible for improved set of fruit, better rooting of cuttings, defoliation prior to harvest, and a number of other useful plant growth responses. Now, some compounds—many still in the experimental stage—have been found to enhance yield, possibly through reduction of the rate of plant respiration or improvement of the efficiency of water utilization. Like true pesticides, these chemicals are not utilized beyond experimental plots until food plant residues have been determined and toxicological studies completed, and until regulatory officials have issued product registrations for specific uses.

It is also important to understand that, when more than one agricultural use is approved, the combined tolerances, considering an average diet, will never exceed the safe daily intake—a figure based on toxicological data with a wide margin of safety, usually 100 to 1, built into it. Such margins are intended to assure safety in spite of variations in diet and in the sensitivity of man, as compared with laboratory animals, and despite occasional higher-than-approved tolerances in food samples. As explained in Chapter 29, subsequent to the 1958 adoption of the Delaney Clause of the U.S. law regulating foods, no pesticide use resulting in a residue in food has been approved for a compound found to be carcinogenic in laboratory tests.

The government of Great Britain assures pesticide safety through operations of the Pesticides Safety Precautions Scheme operated by the Ministry of Agriculture, Fisheries and Food. Although administrative procedures differ from those in the United States, the objective and the nature of premarketing test requirements are similar. The British system of stepwise approval through Trials Clearance, Limited Clearance, Provisional Commercial Clearance, and Full Commercial Clearance has provided public and environmental protection and, at the same time, encouraged innovation on the part of research groups seeking more effective and safer pesticides. In Britain, the separate but coordinated Agricultural Chemicals Approval Scheme is concerned primarily with efficacy and sound use directions. Approval for a new chemical or a new use cannot be given unless safety in use has first been certified under the Pesticides Safety Precautions Scheme.

Canadian pesticide regulatory procedures have similarities to the U.S. system. In reviewing data submitted in support of a pesticide registration application, Canadian authorities are noted for prompt decisions based on the scientific evidence at hand. The European nations require premarketing registration and approval for pest control products, and most countries elsewhere have, or are in the process of establishing, regulatory procedures. Although pesticide registration requirements and approaches to control vary from country to country, they have a common objective of human health protection and an acceptable level of safety to nontarget organisms.

There is worldwide recognition of the need for toxicological information in order to determine safe human exposure to a pesticide by ingestion or other mode of contact. There is also general agreement on the importance of residue determinations in agricultural products following proposed treatment protocols as a basis for decisions on safe uses. Differences in approval of a given pesticide are sometimes related to varying interpretation of toxicological data. Divergent food crops and pest problems in different parts of the world, together with dissimilar pesticide benefits, often contribute to varying actions on the part of regulatory officials. For example, DDT, removed from the market in the United States largely because of alleged adverse effects on nontarget organisms, is still approved for the control of mosquitoes in a number of tropical countries where malaria is a major threat to health.

Watch a ship being unloaded of coffee or bananas and you may wonder how food safety precautions in this country dovetail with the pesticide approval scheme in the country of origin. Do they use the same crop protection chemicals, and do they have the same tolerances in the event that residues are present in the agricultural commodity? Then watch ship after ship being loaded with corn, wheat or soybeans at Vancouver, Houston, New Orleans or Thunder Bay bound for Japan and Europe or other parts of the globe and you may wonder whether food inspection people in the importing countries insist upon the same tolerances for possible residues as those established in Ottawa and Washington. Obviously, the importing country has the last word and can reject any shipment of food containing residues of a chemical at unapproved levels.

One can argue that every country is entitled to its own standards, but the fact cannot be ignored that international trade would be greatly facilitated if all could agree on safe daily intake levels of traces of those chemicals that may be present as a result of approved production or storage practices. If such agreement could then be translated to uniform residue tolerances in major farm commodities, the world of tomorrow would be better nourished as a result of one less impediment to the free movement of foodstuffs in international commerce.

Most food must be grown at home in most countries of the world, but some are past the point, populationwise, where their own agricultural resources are enough to suffice (Japan, for example) and trade is essential to their survival. Then there are the bad years—in both technically advanced countries and developing areas—where limited reserves cannot tide them over and imports of food are essential. Trade can also have land-conservation benefits by permitting countries to raise what they can produce most efficiently.

Appreciable international communication and cooperation

relative to pesticides has evolved, in recent years, but there is a long way to go, and every country—including those with long-established regulatory systems—must be prepared for give and take in the interpretation of toxicological data and the establishment of tolerances. The World Health Organization (WHO) and the Food and Agriculture Organization (FAO) of the United Nations are currently striving for agreement on maximum residue limits for traces of those crop protection chemicals used in many producing areas. Harmonization could help assure that food can be freely exchanged in international commerce of the future. The European Economic Community (EEC) is progressing with standardization of residue tolerances and approved analytical methods. The world's dual need for adequate food—but, at the same time, safe food—requires a continuation of the ever increasing degree of international co-operation that has been evolving in recent years.

# 32

## Some Pesticide Controversies

Just as technical advances sometimes bring forth new hazards, real and perceived, they can also yield new benefits. Trade-offs must be considered and decisions made as to how such technology shall be used and what regulations may be needed to assure a favorable risk/benefit ratio. Controversies seem inevitable during the early years of a new technology as we struggle to evaluate both hazards and benefits. Fear of the unknown sometimes magnifies concerns. There was great public suspicion of electricity early in its domestic use, and in the White House, for a while—after the first electric lights were installed—an electrician is reported to have been employed to turn the switches on and off.

Pesticides constitute a relatively new technology which, all thoughtful people agree, should be integrated with biological, cultural, and genetic controls as discussed in Part I. A truly holistic approach to a consideration of pesticide trade-offs must include *the cost of not employing them*, the cost both environmentally and economically, as well as from the standpoint of public health and nutrition. Obviously, there are two sides to the pesticide question, and both must be brought in focus before decisions on trade-offs that will assure the greatest net benefit can be judged. It should be equally obvious that no broad generalizations on pesticide risks or benefits are justified.

Each compound and each use must be evaluated independently as is the practice in regulatory programs and approval schemes.

Premarketing research on the potential human and environmental hazards of pesticides, as required by regulatory laws and edicts, have gone a long way toward minimizing risks. As with other new technology, projections of potential benefits must be confirmed by field experience. In considering trade-offs, we are often dealing with a moving target. Regulations and technical improvements that reduce risks and newly accumulated data on benefits, as well as possible hazards not projected by initial research, must be constantly fed into the equation.

Assuming that man must protect himself from pests, we must face the pest control controversies and must understand both sides of the issues before we can make wise decisions on trade-offs. Let's review some of these controversies as a part of the background needed for sound decisions in the future.

*Nontarget Organisms.* Critics point out that some pesticides—particularly organochlorine insecticides, including DDT (also called chlorinated hydrocarbons)—have been found to have adverse effects on nontarget species, not only in the treated area but also at a considerable distance, particularly with migrating birds in aquatic environments. Aquatic invertebrates are known to have been adversely affected, and the normal reproduction of some species of birds and fish impaired. Food chains are thus believed to have been interrupted, sometimes, to the detriment of species not directly affected by the pesticidal compound. Defenders of modern pest control chemicals point out that such instances occurred largely as a result of products or application protocols no longer in use in many countries. They further claim that broad generalizations regarding the environmental effects of these now little-used insecticides were often not justified by the facts, and that many uses did not result in appreciable damage to nontarget species. As an example, it is stated that the statistics of catches of many marine species—including fish, shrimp, and crabs—do not indicate population reductions, even during and immediately following the period of greatest use of the organochlorine insecticides. The absence of a reduction in most bird populations is also cited as evidence that pesticide effects on avian species was never as great as often stated.

Defenders question the validity of some of the widely publicized data on the effect of DDT on the thickness of eggshells under natural conditions and claim there have been unwarranted generalizations following controlled experiments. They believe that declines of some bird populations attributed to insecticides were due to other causes. As evidence, they cite the recovery of some populations since DDT was banned followed by a new decline. It is pointed out that wild populations are always having their "ups and downs" and believed that DDT and related compounds were blamed for the "downs" without adequate consideration of other factors and without questioning of the reason for the "ups."

It is pointed out by pesticide defenders that current registration procedures require premarketing studies on possible effects of new pesticidal compounds on representative organisms, as well as a determination of persistence and mobility in the environment. Products presently in wide use are believed not to present a general hazard to organisms off treated areas. It is further stressed that many classes of pesticides, including herbicides, fungicides, and nematicides, have rarely had an adverse effect on an important nontarget wildlife species.

Bees have been killed by inadvertent exposure to insecticides applied to properties within flight-range of their hives. This has created problems for honey producers and, sometimes, for orchardists or other farmers dependent on bees for pollination. Such unfortunate incidents, which most often occur in areas of intensive cropping adjacent to honey producers, are soon recognized, and steps quickly taken to correct them.

Pesticide users recognize these problems and say they are striving for a solution. Much effort is being made in honey-producing areas to utilize insecticides of low residuality and to organize communications so beekeepers will be made aware of crop treatment plans and cover their hives for short periods after nearby fields or orchards have been given an insecticidal application. It is often pointed out that production of honey has not declined during the past four decades—which covers the period of growth in the use of farm chemicals from a low to a high level. There are, of course, year to year variations in honey production per hive, but these appear to be more related to moisture conditions than to other factors; for example, the reduced yields experienced in the western United States during the dry years of the mid-1970s.

*Insect Predators and Parasites.* Critics emphasize that insecticidal applications often result in a reduction of beneficial species which prey on the pest being controlled. Consequently, the target species may have an even greater resurgence following treatment. It is further pointed out that some potentially harmful secondary species normally kept in check by natural predators have a better chance of reaching damaging populations as a result of certain insecticidal treatments. Thus, a species not normally a pest can become one. As a classic example, critics cite the increase in European red mite following the wide-scale orchard use of DDT, thus necessitating steps to manage mite populations.

Farmers and others on the "pest control firing line" agree that these adverse effects of insecticides have sometimes occurred and that efforts at minimizing the reduction of populations of desirable predators and parasites are imperative. It is argued that, with ever expanding knowledge of the biology of pest species and improved population-monitoring techniques, timing and dosages are increasingly being chosen with a "rifle" rather than a "shotgun" approach. With the costs of insecticides inflating rapidly, users say that they have every reason to take advantage of natural predators and parasites to the fullest extent possible.

Pesticide users, striving to apply the best integrated management techniques available, wish they had more specific compounds, ones that would control the target species with the least possible adverse effect on desirable predators and parasites. They complain that picayunish regulations and long delays in approval have so increased the cost of developing specialized products that many highly specific chemical pesticides—as well as biological agents already discovered—are not moving toward commercialization.* No one argues that, even with more specific chemicals, the reduction of predatory species below their effective level can always be avoided. Trade-offs will have to be considered if pest control decisions of the future are to be wisely made.

*Bioconcentration and biomagnification.* The fact that certain organochlorine insecticides are known to accumulate or bioconcentrate in some organisms to a level higher than found in the surrounding environment is not disputed. For example, fish and nonvertebrates living in water that contains a small fraction of a part per million of DDT have been known to accumulate several parts per million of this insecticide. Persistent compounds readily soluble in fat and having a slow rate of metabolism in biological systems are the ones most likely to be bioconcentrated. The phenomenon does not proceed indefinitely, however. Whether the pesticide in question is accumulated only from its surroundings—water, in the case of aquatic life—or also from its food, each species appears to reach an equilibrium concentration the level of which depends on many factors, including rates of metabolism, defecation, and excretion, as well as various environmental influences. Terrestrial vertebrates can also accumulate persistent and fat-soluble compounds, but largely through their food.

Aside from bioconcentration, it has frequently been stated

_____

*\*Of Mites and Men* by William Tucker, Harpers (magazine), August 1978.

that certain pesticides are biomagnified, that they reach higher and higher levels with each successive step in a food chain. For example, a booklet published by the Environmental Protection Agency in the United States, entitled *Pesticide Registration: How it Protects You, Your Family and the Environment*, refers to chemicals that:

> ... can move up in the food chain. For example: small amounts of chemicals absorbed by plankton and insects are transferred in increasing concentrations to fish, birds and animals and eventually to man through food.

The concept of biomagnification has been widely held since propounded by Rachel Carson in *Silent Spring* in 1962. Indeed, the concentrations of certain pesticides found in organisms at different levels in food chains have sometimes appeared to confirm its occurrence. But many are now convinced that the concept of biomagnification has been greatly oversimplified and that analytical data on which the theory was based has often been misread.

A number of scientists who have devoted their research efforts to studies of the fate of pesticides that may escape into the environment now believe that each species of animal life accumulates various foreign substances only up to a "plateau" level. This appears to be true regardless of whether these substances are derived from their food or their surroundings— water, in the case of aquatic species. It is now known that certain fish high on an aquatic food chain can contain as great a residue of a chemical when derived solely from water as when they also consume it in their food. It is pointed out that tiny organisms low on the food chain, such as unicellular algae, often contain as much of certain pesticides per unit of body weight as organisms high on the chain. They tend to adsorb many substances on their surface, and—because of their small size—they have a large surface-to-weight ratio. Therefore, concentrations of pesticides in these plankton are often as high or higher than in other organisms up the chain.

Birds and other terrestrial animals have at times been injured by certain pesticides derived from their food, and relatively high subacute levels have been found on chemical analysis of sacrificed specimens. This phenomenon was observed most often with the now little-used organochlorine insecticides. It is currently believed by many researchers specializing in studies on the fate of chemicals in the environment that the amount accumulated depends not so much on the quantity present in food as on the capacity of various species to accumulate certain compounds. There is evidence that certain organisms consuming large amounts of food containing minute amounts of certain pollutants can concentrate them, while other species with a lower retention capability end up with lower levels than existed in their food.

Those who question the scientific validity of the popular generalization about the occurrence of biomagnification, point out that there has frequently been mistaken compound identification and other misinterpretation of data related to the amount of certain pesticides found upon chemical analysis of species at different levels of a food chain. For example, "wet-weight" of samples of food have sometimes been compared with "dry-weight" of tissues of birds and other animals higher on the chain. These specialists do not say that biomagnification never occurs, but they do contend that pesticide critics have generalized from rare occurrences—unwittingly, or with intent to frighten the public.*

*Spray Schedules and IPM.* Critics say that environmental pollution with agricultural pesticides and adverse effects on beneficial species result from unnecessary applications. They

---

*George Klaus and Karen Bolander, *Ecological Sanity* (New York: David McKay Company, Inc., 1977).

claim that most crop producers spray according to the calendar and ignore the latest integrated pest management (IPM) techniques involving population-monitoring followed by pesticide applications only as need indicates. They further complain that fruit and vegetable producers often treat with pesticides just to improve the appearance of their produce: a cosmetic effect. It is claimed that a greater use of pesticides than is actually needed results from advice given growers by pesticide salesmen.

Defenders of agricultural pesticides laugh at the notion that farmers base their decisions largely on advice of salesmen and emphasize the vested interest every producer has in lowering costs wherever possible. All will agree that, in the past, monitoring systems for determining just when an application should be made have been very imprecise. As one grower in Michigan put it, "I used to have to guess when the humidity, temperature, and moisture retention on leaves were just right for apple-scab spore germination and I should start spraying. Now, I get a reading by phone from a computer that plugs all these variables into a calculation that tells me when to apply my scab fungicide."

Growers say they are increasingly in tune with monitoring technology and are adopting new techniques of estimating insect populations and scientifically determining when treatment for plant disease is necessary. They report that many of the programs that may eventually result in less-frequent pesticide applications or possibly lower dosages are still in the research stage and, with their tremendous investment in a crop, they can ill afford to experiment widely on unproven programs. Farm magazines and agricultural extension meetings spread news of the latest IPM methods for different crops, but producers say that proven programs ready for general use are available for only a portion of their pests. They further state that some IPM programs proven effective in one area are useless in another because of different pests or divergent climatic conditions (Chapter 8).

Application by the calendar is still necessary for many pests, according to producers, as well as many research and extension specialists—for example, for soil pests that must be treated before planting. If serious damage is almost certain to occur from nematodes or certain soil insects, there is little choice but to apply the appropriate pesticide before planting. Rotations are sometimes helpful (Chapter 5), but growers point out that they are not always practical and do not always succeed in providing adequate control. In the case of preemergence herbicides, it is claimed that a grower knows pretty well what weed seeds are in his soil and that he must often routinely apply an appropriate preemergence product before the crop and weeds come up.

The oft-repeated criticism that growers use pesticides for "cosmetic" purposes is countered with a reminder that, even if consumers would accept less attractive produce than is now available, several fruit and vegetable pests that detract from appearance would also cause poor storage life, apple scab, and various fruit-infesting insects, among other things. The control of rust mites for improved appearance of citrus is often cited as an example of pest control for cosmetic reasons. Growers agree that productivity may not be impaired by modest populations, but they point out that young fruit may be damaged and drop prematurely if this mite species is allowed to go unchecked. One grower asked, "Even though rust mites only affected fruit appearance, what environmental or human health threat is associated with the use of currently approved miticides that would justify the loss of a market for the crop of an individual or a whole industry because of unattractive fruit?"

*Resistance to Pesticides.* Tolerance to pest control chemicals is often developed by insects and mites and, sometimes, by fungi and other classes of organisms following extended use.

Critics cite this phenomenon as an argument against the employment of pesticides. They say an increased dosage is often needed to achieve a given level of suppression and that resistance is sometimes intensified to the point that a chemical becomes worthless for the control of certain species.

Those relying on pesticides as one of their "tools" in defense against economic losses and risks to health are fully aware—indeed often painfully aware—of the facts of resistance. But they believe that pesticides should be used in management programs, with recognition that, sooner or later, different chemicals or methods may have to be employed. Some users alternate different pesticides, in the belief that the development of resistance is thereby delayed, if not avoided (Chapter 18). They say that the genetic modification that accounts for resistance is not universal; it sometimes occurs only in restricted geographic areas.

They further emphasize that, to certain chemicals, only a modest degree of resistance has evolved, while others are fully as effective after years of use as they were when first discovered. Certain mite control compounds are cited as examples of this variability. With their numerous generations within a single year, mites have been known to become resistant to some chemicals within a very short time, while to one leading miticide, no resistance is known to have appeared, even after many years of use. One objective of researchers seeking improved pesticides is to design molecules that the pest will have difficulty in becoming resistant to. It is further stated that relatively little resistance has evolved to many herbicides, fumigants, and even some fungicides.

Pesticide defenders ask why critics make wholesale condemnations of pest control chemicals because of the development of various degrees of tolerance while praising resistant crop varieties—which so often lose their resistance as a result of genetic adaptation on the part of the pest. These defenders believe that man must employ all the "tools" available to him, if he is to avoid losses he can ill afford, and that his defenses must include both resistant varieties and pesticides, even though the weaknesses of both approaches are well recognized.

*Energy Conservation.* Those against the use of pesticides often argue that most chemicals are derived from petroleum, an energy resource we should be conserving. Defenders say we gain far more than we expend; they say that increased yields as a result of good crop protection mean less total energy must be spent for each unit of production. One example is the half-gallon or less of petroleum equivalent required to make, package, and apply the 2,4-D or MCPA employed for weed control in small grains—a practice which often results in an increase of 5 bushels or more per acre (Chapter 19). A half-gallon of gasoline will no more than transport many food shoppers to the store and back, while 5 bushels of wheat will make the flour required for about 300 one-pound loaves of bread.

In reality, the energy required for the manufacture and distribution of pesticides constitutes only a modest portion of the total energy input required for most food crops. Assuming these crops are to be produced by one means or another, the manufacturing energy required for essential pesticides must be weighed against the total energy requirements of the increased acreage needed to make up for pest losses if good crop protection is not achieved. Referring back to the preceding paragraph, proponents of selective herbicides point out that to increase total production of wheat or another bread grain by five bushels merely by expanding the area planted without the benefit of selective herbicides would require an energy input far greater than a half-gallon of petroleum.

Agriculturists believe that the very top priority for petroleum, when a tight supply situation develops, should be for the essential production inputs of basic food producers and of those who transform and transport farm commodities to food products at local markets. They think that, after experiencing a few empty shelves at the supermarket, all would agree. (Some of the energy-conservation benefits of pesticides were discussed in Chapter 17.)

*Soil Fertility.* Critics claiming special concern for soil fertility say that earthworms have been killed by certain pesticide treatments, and the fear is expressed that other desirable soil organisms are also adversely affected. Defenders counter that the residual compounds that have at times been suspected of significantly reducing the populations of desirable soil-inhabiting organisms are primarily the organochlorine compounds, little used today. They also claim that the experiments now required to define possible adverse environmental effects show that, almost without exception, pesticides which reach the soil following application to growing crops have no more than a partial and temporary effect on soil organisms. Most soil-applied pesticides are laid down in bands: therefore repopulation from the untreated strips soon takes place, if organisms are killed in the treated zone. Even where a larger area of soil is treated with a fumigant that kills earthworms completely, reinvasion from surrounding land is observed to occur rather quickly.

Pesticide defenders cite the accumulation of organic matter from the residues of high-yielding crops as an answer to the belief of some organic gardening enthusiasts that, somehow, chemicals are destroying fertility by "poisoning" the soil.* It is argued that, with few exceptions, modern pesticides degrade into compounds identical to those found in nature—and ultimately to carbon dioxide and water. Defenders further cite the

---

*For a further discussion of soil fertility as related to modern agriculture see Chapters 3, 5, and 6 of *The Food in Your Future* by Keith C. Barrons (New York: VanNostrand Reinhold Company, 1975).

soil-conserving benefits of herbicides that make possible reduced tillage of land subject to erosion (Chapter 17).

*Pest Losses.* It has sometimes been claimed by critics that agricultural losses from pests are as great or greater now than before pesticides became widely used. In the United States, these allegations are claimed to be based on loss estimates made by the U.S. Department of Agriculture.

Pesticide users question the validity of such estimates, declaring that, today, we are aware of losses from a wide range of pests that were little recognized a few decades ago—for example, from nematodes and other soil organisms, stalk rots, leaf hoppers, and many viruses. They believe that, had earlier crop loss estimates been made on the same basis as those of today, such loss estimates would have been far higher than the old estimates referred to.

Pesticide defenders believe that the important measurement is the volume of food saved and the yield increases realized through modern crop protection, not someone's estimate of percentages lost due to pests. The phenomenal yield increases of recent decades resulted from many factors, including:

> superior breeding;
> better plant nutrition;
> improved cultural practices;
> a higher level of crop protection.

Marked increases in crop yields are by no means confined to the United States, but experience in this country, as indicated by records of the U.S. Department of Agriculture, serves to illustrate what agriculture has done to assure us of adequate food during recent decades. A fifty-year history of potato yields was presented in Chapter 10 and a full century of corn yields in Chapter 24. The following table gives a five-decade summary of average yields of other major grain crops.

| Average yield of some major U.S. grain crops in bushels per acre for decade beginning — | | | | |
|---|---|---|---|---|
| | 1930 | 1940 | 1950 | 1960 | 1970 |
| Wheat | 13 | 17 | 20 | 26 | 31 |
| Barley | 21 | 25 | 29 | 38 | 44 |
| Sorghum | 13 | 18 | 24 | 48 | 54 |
| Oats | 27 | 33 | 36 | 47 | 51 |
| Soybeans | 16 | 19 | 21 | 25 | 28 |

During the last half-century, U.S. forage crops have also increased substantially in yield, silage corn from 6.3 tons per acre in the 1930s to 12.2 tons in the 1970s, and hay from 1.2 tons to 2.1 tons during the same time span. Cotton, which yielded only 205 pounds of fiber per acre during the 1930s, reached an average of 476 pounds during the decade of the '70s.

Farmers who assure our food abundance by employing the inputs necessary for high yields consider crop protection chemicals important to their success. It is often pointed out that the crop yield revolution beginning in the 1940s coincided with the discovery and subsequent commercialization of more effective pest control agents.

Pesticide usage was relatively modest during the decades of the 1930s and 1940s, but increased markedly in the 1950s and 1960s. Further increases in the volume of pesticides used during the 1970s was largely due to wider application of herbicides for selective weed control in major crops. A significant portion of the increases in productivity per acre are believed, by pesticide defenders, to be the result of the use of crop protection chemicals. They state that, without effective control, losses of well-fertilized and potentially high-yielding crops would often be disastrous and that, at this time, chemicals are an important component of the majority of effective pest management programs.

*Volume of Use of Pesticides.* Critics often decry the fact that so many millions of pounds of pest control chemicals are employed around the world. They tend to equate volume with adverse human or environmental effects, regardless of the nature of the use or the benefits derived. Pesticide users hope they can someday protect crops and livestock and accomplish other pest management objectives with fewer pounds of chemicals at a lower cost than at present, but reject the notion that pesticides are "bad, per se." Users maintain that, when properly chosen and applied, pest control chemicals have a positive net benefit to mankind and that there are many situations where more pounds, rather than fewer, would improve the environment—for example, to reduce rodent and insect damage to grain and other stored products for which we have already devoted land and energy resources. A further example is the reduction of nitrogen fertilizer losses from the soil by the slowing down of the nitrification process through the use of an inhibitor (Chapter 19).

On the other side of the volume-of-use coin is the fact that many newer pesticides require a lower rate of application than older ones, an important consideration in possible human—or wildlife—exposure incidental to use. A compound with a higher level of toxicity as indicated by measurements, such as the $LD_{50}$ (Chapter 29), may actually present a lower practical hazard than older less-toxic materials if the rate of application of the new compound is significantly less. The grounds of the Florida condominium where we live for part of the year receive a modern chinch bug insecticide for protection of the lawn at about 1 ounce per 1000 square feet, whereas an older material had to be applied at 3 ounces per 1000. As the new material protects over a longer period, the total amount of insecticide applied per year is reduced significantly.

# 33

## The Human Health Question

Recently I asked a sampling of individuals what first came to mind at the mention of pesticide safety to people. Many thought of their desire to avoid harmful residues in food, while one wondered about the safety of a cockroach control treatment being applied periodically to her home by a pest control contractor. A vegetable grower reflected on the precautions he should take when spraying an insecticide and the time required following application before it was safe to harvest. A worker in a pesticide formulating plant said that his safety mask and instructions for washing up after work first came to mind, while an agricultural extension agent thought of the precautionary statements on pesticide labels. My sample of interviewees was hardly adequate to justify generalizations, but it did point out that pesticide safety is a many-sided matter—and often as controversial as questions of environmental safety discussed in the preceding chapter.

*Manufacturing and Distribution.* To start at the beginning, let's consider the plant workers' concern: safety in manufacturing pesticide products. Today, active ingredients are usually manufactured in a large chemical complex with a high degree of control of possible atmospheric contamination and appropriate treatment of waste products to avoid pollution of water. Workers generally operate under strict rules of industrial hygiene. Standards for exposure limits are set following extensive toxicological research, the same tests as needed to define appropriate precautions for pesticide applicators.

Remarkably few instances of injury to workers in those chemical plants manufacturing pesticides are known to have occurred. Several studies of workers with a long history of exposure to DDT in plants where this pesticide was manufactured failed to indicate unusual health problems. One compound, DBCP, used as a nematicidal fumigant, was found after several years of production to induce a lowered sperm count in some men exposed during the manufacturing process. No other untoward human effects were noted. Whether standards for tolerable exposure were set too high or there was carelessness in handling the product is not known.

Faulty engineering has occasionally resulted in pesticide plant explosions, with injury or loss of life to workers. The past record of the chemical industry in safely disposing of wastes from manufacturing operations has, at times, been unacceptable—example: pollution of the James River in the United States by wastes from manufacture of an insecticide formerly used for fire ant control. Present regulations requiring that manufacturers account for toxic wastes and dispose of them safely have gone a long way toward eliminating potential hazards. Dusts and vapors from pesticide formulating plants have been of concern, but are now generally controlled through the use of technically advanced equipment. Workers are well protected through appropriate clothing, and masks where needed.

In recent years, distribution of pesticides has been largely in the final package going to the user, and transportation accidents have been rare. The few which have occurred emphasize the care needed—particularly in bulk handling on highways or railroads—for all products that can cause harm to the public or to the environment.

*Safety in Pesticide Application.* Consideration must be

given, not only to persons actually applying the pesticide, but also to others who might contact a hazardous material if appropriate precautions were not taken—for example, to crop harvesting crews if a field is entered too soon following treatment or to people in the vicinity at the time of application that could be exposed to spray or dust drift. All these are factored into regulatory decisions concerned with classification of products for general or restricted use and in the preparation of precautionary statements on labels. Further consideration must be given to safe disposal of pesticide containers, and—where a significant hazard may be involved—specific disposal directions are now provided as a part of the label precautions. Instances of poisoning have occurred when pesticide containers, improperly cleaned, were used for transporting drinking water. Current regulations which fix responsibility for appropriate container isolation and disposal should materially reduce this type of accident.

Instances of carelessness have resulted in illness or even deaths to applicators; such occurrences have been rare, however, in recent years. With the heightening of applicator awareness of potential hazards, aided by current systems of product classification and appropriate certification of farmers and other applicators before they can purchase and apply a product in the restricted category, the incidence of adverse effects appears to be declining. For products with a potential for injury to harvesting crews or others who may be involved with farm work in a pesticide-treated area, safe re-entry time is determined by research in advance of approval for a given use. The presence of residues that workers might be exposed to at various post-treatment times indicates how long a waiting period will be specified on the label.

It has long been recognized that compounds with a potential for harm at low concentrations must be applied with extra care to keep spraydrift at a low level, whether ground rigs or aircraft are involved. Special aircraft-wing, spray-boom, and nozzle designs, plus unique formulations and rigorous attention to meteriological conditions are combining to minimize the tiny droplets subject to drifting off treated areas.

*Food Residues.* As discussed in Chapter 29 on toxicology and Chapter 31 on regulation, there has long been concern for possible hazard to the public from traces of pesticides that may be present in foods. Allowable tolerances in a given food as established by governmental agencies in various countries depend on the best estimate toxicologists can make of safe intakes and the importance of the food in question in the diet. Wide margins of safety are provided to allow for 1) variations in intake, 2) occasional higher levels of residues than permitted by legal tolerances, and 3) the possibility that man may be more sensitive to the compound in question than the laboratory animals used in toxicological tests.

There is sometimes criticism of governmental agencies for inadequate policing of foods for possible residues. Such criticism is usually based on the fact that sampling of foods from the marketplace followed by chemical analysis sometimes reveals that a small proportion shows a residue moderately higher than tolerance limits. This does not necessarily indicate a danger to human health; there is a wide margin of safety provided for in establishing tolerances. Further, if an occasional sample exceeds official tolerances, it does not necessarily mean that a food producer has failed to follow directions. Variability is common in biological phenomena—one reason for allowing wide safety margins in establishing legal limits. Nonetheless, tolerance excesses must not be taken lightly, and if they occur frequently in a given food, label adequacy must be reviewed and producer compliance investigated.

Pesticide residues in imported food have sometimes been of concern, particularly with fresh fruits and vegetables, where their perishable nature and frequency of shipment can strain the receiving country's sampling and analytical capabilities. Occasionally, a pesticide used in the production of an export crop has not been approved by the receiving country, either because of diverse interpretation of toxicological data or because the manufacturer has not yet submitted an application. Even though the residues in question may be within tolerances eventually approved by the importing country, exporters have a responsibility to abide by the regulations and tolerances of those to whom they ship foodstuffs. The banning or confiscation of imports can be a severe penalty for not living up to this responsibility. Countries and growers producing especially for export must understand the residue tolerances allowed in countries of destination and must adjust pest control programs accordingly. Efforts are under way to minimize these problems of international trade by establishing uniform tolerances (Chapter 31).

*The Fear of Cancer.* Pesticide risks to factory employes, to applicators, and to farm workers are most often perceived to be of an acute nature—an immediate illness as a result of excessive exposure. The public, on the other hand, with little likelihood of frequent or direct contact and rare instances of acute illness or injury, is more concerned with the possibility of cancer or other chronic effects as a result of traces in food.

Even though the word *cancer* has long been among the most feared in the language and, until a few years ago, was used only in hushed tones, little thought was given to its causes. As some families seemed to be prone to the disease, heredity was assumed to play a dominant role in its occurrence. Many considered it "God's will," something that happened that man could do nothing about. Although radiation from radium and other sources had been incriminated and a few occupational relationships to the incidence of cancer were known (e.g., chimney sweeps and workers with certain dyes), little thought was given to the influence of the world around us on the occurrence of this dread disease.

Then, during the 1950s, numerous investigations showed that cigarette smoke was a prime causative factor in lung cancer. It was also established that workers in asbestos plants and others with high exposure to this substance had a disproportionately greater chance of developing lung cancer. Suddenly all of our surroundings were suspect: the air we breathe, our food, our drink, and all kinds of substances to which one might be exposed in the workplace or at home. The great hunt for carcinogens was on. We wanted to forget the old notion of heredity as a prime factor. It was much pleasanter to dwell upon possible culprits that we could eliminate from things we ingest or from the environment around us.

In the United States, one legal consequence of the new awareness that specific substances may increase the incidence of cancer took the form of the Delaney Clause of the law regulating the purity of foods. As discussed in Chapter 29, this clause in the amendments passed by the Congress in 1958 mandates that no tolerances can be established for any substance added to food that is found to be carcinogenic in man or animals. Here are the exact words:

> No additive shall be deemed to be safe if it is found to induce cancer when ingested by man or animal, or if it is found, after tests which are appropriate for the evaluation of the safety of food additives, to induce cancer in man or animals.

Although traces of pesticides that may be present in foods are not intentional additives, but rather incidental to production or storage, the Delaney Clause has served as a guide to tolerance establishment. No proposed pesticide use has been approved which may result in a residue of any material that has shown evidence of carcinogenicity in laboratory tests with rats or other animals.

The search for carcinogens in recent years has encompassed all manner of substances, including air and water pollutants, cosmetics, and household products, as well as those found in the workplace. Much emphasis has been placed on

food additives intentionally used in processing and also incidental residues, including pesticides. The premarketing safety tests
that had been required for pesticides introduced prior to passage
of the Delaney Clause did not place major emphasis on carcino-
genesis, so a wide range of retesting was instituted using mice,
rats, dogs, and occasionally primates. The dosage range con-
cepts described in Chapter 29 came into wide use in designing
protocols for retesting older products, as well as for new ones
being considered for commercialization.

*The Cranberry Episode.* The first pesticide use to be "outlawed" under the new regulations involved aminotrizole, an
herbicide employed primarily for the control of vegetation
along roadsides, but also on certain food crops. Many readers
will recall the cranberry episode in the United States in the fall
of 1959. Tests had shown that this compound increased the incidence of thyroid tumors in rats when administered in high
doses in dietary feeding experiments. The U.S. Department of
Agriculture, then responsible for pesticide regulation, believed
there was no choice but to deny registration of the product for
food crop use. When it was found by the Food and Drug Administration that some cranberries contained identifiable traces
of aminotrizole, that agency was convinced it was obligated by
law to confiscate huge quantities of this fruit.

Cranberry growers and many scientists claimed that federal
regulatory officials were being unreasonable in their interpretation of the law, and that the tests with rats on which their decisions were based were inappropriate because of the high levels
employed and thus not in accord with the Delaney Clause,
which called for "tests which are appropriate for the evaluation
of the safety of food additives." It was pointed out that, if a
person ate 20 grams of cranberries daily and these berries contained as much as .5 ppm of aminotriazole, it would be equivalent to about .01 ppm in the total diet, assuming a food intake of only 1000 grams daily. The lowest dose to show a
thyroid effect (and that on only an occasional animal) was

10 ppm in their total diet, one thousand times greater than the above projected human intake. It was pointed out that many lots of berries contained lower levels of the contaminant and that few people eat cranberry products daily.

Defenders of the Delaney concept contended that there is no threshold for activity of carcinogens, and a compound that appeared to induce tumors in animals at any dosage might cause cancer in an occasional human, even at an infinitesimally small dose. So the great debate was joined—the threshold versus the no threshold for carcinogenicity camps. The controversy continues, because neither side can prove its point, as discussed in Chapter 29.

*The DDT Conflict.* During the decade before and after passage of the Delaney Amendment, millions of lives were saved in many parts of the world and millions of serious illnesses prevented through the control of malaria-transmitting mosquitoes with DDT. For centuries, malaria has been the great killer disease in most tropical areas. (As indicated in Chapter 20, DDT is credited with having averted a potentially devastating outbreak of typhus in southern Europe during World War II).

But while the benefits of DDT in terms of better health and improved food production were being reaped, its possible adverse effects were beginning to be more clearly identified. The concerns expressd in Rachel Carson's book *Silent Spring*, published in 1962, triggered widespread anxiety regarding the negative side of the pesticide matter.

As discussed in Chapter 32, the persistence and mobility of DDT—together with alleged effects on nontarget organisms, particularly on the fertility of certain birds and fish—prompted widespread demands for banning the use of this compound. Defenders pointed out its importance in vector control, in the production of a number of specialized crops, and in forest insect control, in addition to its major use on cotton. Their claims that effective substitutes for several crop uses were not available, together with the argument that much of the DDT reaching streams and lakes was derived from use in municipalities for spraying street trees, failed to quell demands for outlawing it for all uses, including its use on specialized crops.

Controversy over the continuing use of DDT intensified, in the late 1960s, and the Environmental Protection Agency—which in 1970 took over regulatory controls governing pesticide registration and use from the Department of Agriculture—was pressured by groups with environmental concerns to cancel all approvals for use of this pesticide. An advisory committee review and numerous legal maneuvers by both sides were followed by a formal hearing by a U.S. Government Examiner, a quasi-judicial affair conducted under legal rules of evidence and entailing examination and cross examination of witnesses.

The generally recognized low level of toxicity of DDT to humans confirmed by studies of long-time workers in chemical plants where the compound was made, as well as the records of applicators in daily contact with it over a period of many years, was countered by some researchers who claimed it was a carcinogen. Defenders documented several experiments in which no evidence of an increase in tumors in laboratory animals was revealed, but others claimed that two or three experiments with mice established carcinogenicity. Defenders cited the occurrence of numerous tumors in the control groups of animals receiving no DDT and suggested the inappropriateness of inbred mice, so likely to develop tumors from viruses or other causes, as reliable test animals. The hearing Examiner referred to in the preceding paragraph ruled that DDT is not a carcinogenic hazard to man, nor does it pose mutagenic or teratogenic (birth defect) risks. Today it appears that most cancer researchers and those concerned with other irreversible effects agree with this conclusion. The summary* of a recent world

*Environmental Health Criteria 9: *DDT and its Derivatives.* Published under the Joint sponsorship of the United Nations Environmental Programme and the World Health Organization (Geneva, Switzerland, 1978).

survey by the Environmental Programme and the World Health Organization of the United Nations states that DDT ". . . has had an exceptionally good safety record with no indication of teratogenic or mutagenic effects or of carcinogenicity in man."

The claim that adverse effects on nontarget forms of life outweighed the benefits of DDT was countered by its defenders who pointed out what they interpreted as weaknesses in many investigations and unjustifiable claims regarding the amount of the compound present in the environment. For example, various organisms were found to have a lower level of DDT in their bodies than one would anticipate, based on alleged persistence and bioaccumulation, thus raising questions regarding the widely held belief that the compound was extremely slow to decompose. Some qualified analytical chemists, on reviewing reports of appreciable levels of DDT in soil, water, and various forms of life, stated that many analysts failed to distinguish between this compound and the PCBs—industrial chemicals known to be persistent and to have escaped into the environment. Further, several researchers claimed to have identified naturally occurring chlorinated organic compounds, and the possibilities were emphasized that some analytical data may have given higher readings for DDT than actually existed, because of these chemically related substances.

In spite of recommendations of the Examiner in the hearings referred to, the administrator of the Environmental Protection Agency ruled that practically all registrations for DDT in the United States should be canceled. Amid similar claims and counterclaims, a number of other countries have canceled approvals for major uses of DDT, while some—particularly in areas where malaria remains a serious health threat—continue to permit its use, primarily as a part of the World Health Organization program to mitigate that disease.

*Other Insecticides.* During the last three decades, remarkable advances in analytical chemistry have made it possible to measure fantastically minute levels of pesticide residues that might be present in food. As a consequence, some agricultural products once thought to have no traces of a given pesticide have been shown to carry very low levels, often measured in parts per billion and recently even in parts per trillion. The fact that one part per billion of ten million dollars, ($10,000,000) equals only one cent (1¢) will illustrate what fantastically tiny amounts of substances can be measured by today's skilled analytical chemists using highly sophisticated equipment. The saying that "Today's chemists can find anything in anything" appears to be nearing reality.

While improvements in analytical techniques were being made, new approaches to conducting and interpreting long-term toxicological tests for carcinogenicity have evolved as discussed in Chapter 29. Distinction was formerly made between tumors in laboratory animals that appeared to be benign and those that were clearly carcinogenic. The Environmental Protection Agency now regards any tumor as suspect and any compound that increases the incidence of tumors in laboratory animals as carcinogenic.

Since the controversy over the safety of DDT, several other insecticides of the organochlorine group have likewise had their governmental approvals terminated. As a result of new toxicology data from tests with laboratory animals and given today's interpretation, they were declared to be carcinogenic. With the new analytical technology, traces of some compounds were found in agricultural commodities. Banned compounds include Aldrin and Dieldrin, formerly used for corn rootworm control. Chlordane and heptachlor are being phased out for agricultural uses in the United States, but will continue to be available for termite control. Kepone, myrex, and lindane are other formerly important insecticides the use of which has been eliminated or greatly restricted by new regulations.

Defenders of various insecticides threatened with loss of governmental approval point out their benefits and their low level of acute toxicity to man at exposure levels likely to be encountered. They question the conclusions some have drawn regarding effects on nontarget organisms. But in the United States the Delaney Clause is "the law of the land," so farmers, public health officials, and others concerned with pest management have had to turn to alternate pesticides or other methods of control.

Partly as a consequence of the demise of the organochlorine compounds, two other groups of insecticides—the organophosphates and the carbamates—have come into wide use. Both tend to be considerably less persistent in the environment than the organochlorines, and—although some are more acutely toxic to mammals—they are considered safe from both the human health and the environmental standpoint when employed according to label directions and precautionary statements. Those that require extra care in handling have been placed in the restricted classification in the United States.

Other insecticides now in use include the synthetic pyrethroids and also extracted pyrethrins, as well as some other naturally occurring compounds. New chemicals with unusual modes of action, including pheromones and hormones, are beginning to be employed or are awaiting approval. Insecticide products now available in the United States and most other countries have been carefully researched for safety, and some are under further review. As with any human activity, it is impossible to prove that there is absolutely no risk involved in the use of a given pesticide. Defenders argue that final decisions must be made in part by balancing the risk some believe may exist with known benefits.

*The Imported Fire Ant.* Few pest control programs have been more controversial than that involving the imported fire ant in the southeastern United States. Domestic members of the fire ant group have never been significant pests, but a South American species that gained entry many years ago spread rapidly throughout the Gulf Coast states. Northward extension of its range appears to be limited by intolerance to cold winter weather.

The fire ant is not a major pest of crops or livestock, but its mounds may interfere with the operation of mowers and other farm machinery. There is disagreement regarding agricultural effects; however, some agronomists believe that pasture and crop damage may be significant. On the other hand, fire ants are known to be predators and, at times, probably help keep destructive insects in check. The fire ant's sting discourages laborers from working in heavily infested fields; most of those stung experience only temporary discomfort, but a few need medical treatment because of a hypersensitive reaction that can be life-threatening. Secondary infection can develop at the site of multiple stings, probably from breaks in the skin caused by scratching.

I was recently reminded of the painful nature of fire ant attack when I received half a dozen stings on my hand while picking up a stray piece of paper on the lawn. There was no mound on the property, but a large one was evident in an adjacent vacant lot. The tiny white festers at the top of each sting disappeared in a day or two, but I could imagine what agony one would be in if one received dozens of them. Probably more people receive stings from ants inhabiting home grounds and recreational areas than roadsides and agricultural lands. Some measure of control is often necessary in such areas, in the interests of health and comfort.

In response to growing public demand and to resolutions by southern agricultural organizations, the U.S. Congress enacted legislation in 1957 providing for joint state-federal fire ant quarantine and eradication programs. Quarantine efforts have had some success, but attempts at eradication are now generally recognized as failures. Some ants inevitably escape any insecticidal application, and a treated area eventually becomes reinfested.

Initial attempts at eradication were with the organochlorine insecticides dieldrin and heptachlor—claimed by many to be harmful to wildlife. With the banning of this class of insecticide a new chemical, effective in a bait formulation, called mirex, was widely employed. Claims were then made by some that mirex, too, presented environmental hazards, and its use was terminated following the interpretation of dietary feeding studies with laboratory animals that the compound is carcinogenic. Many agriculturists and others in the southern United States claimed that this action by the Environmental Protection Agency was entirely unjustified, based on the unlikelihood of mirex being present in food when applied with ground equipment as well as questionable interpretation of the toxicological data.

At present, individual mound treatment with such organophosphorous insecticides as diazinon and chloropyrophos is the recommended chemical approach to control. One new material believed to be safe for aerial application to pastures and other areas where individual mound treatment is impractical is currently under test, and may provide a new weapon for managing this troublesome pest. Hopefully current research on biological agents which are believed to keep fire ant populations within bounds in their native areas of South America will eventually help alleviate the problem elsewhere.

*The 2,4,5-T - Dioxin Controversy.* Among the many commercial pest control chemicals other than insecticides relatively few have come under suspicion of undue human or environmental hazard following initial approval and use. One currently under review is 2,4,5-trichlorophenoxyacetic acid (2,4,5-T) which itself has seldom been suspected of presenting a hazard but which cannot be manufactured without the presence of a trace impurity commonly known as a dioxin or more specifically as TCDD. Actually the dioxins constitute a large family of chemical compounds some of which have appreciable toxicity to animal life. The one formed when trichlorophenol, an intermediate in the synthesis of 2,4,5-T and certain related compounds, is manufactured is commonly referred to as TCDD and has the chemical name: 2,3,7,8-tetrachlorodibenzo-p-dioxin.

The toxicity and possible hazard to humans or other forms of life from this contaminant first became of concern during the Viet Nam War when 2,4,5-T was a component of a U.S. military formulation known as Agent Orange used for the defoliation of trees that provided protective cover for enemy troops and military equipment. This agent which was sprayed in Viet Nam at levels far higher than its components are employed in non warfare applications has never been used for agricultural, forestry or right-of-way vegetation control.

As now manufactured 2,4,5-T contains less than .1 part per million of TCDD. Thus if a gallon of one of today's 2,4,5-T herbicide products containing 4 pounds of the active ingredient is applied to an acre of young forest for weed-tree control the amount of TCDD in the spray is no more than one ten millionth of 4 pounds or .00018 grams. Quantitatively this is equivalent to applying a maximum of one 5-grain aspirin tablet or approximately one-third of a gram to 1800 acres or nearly three square miles. A number of uses of 2,4,5-T involve doses lower than those used in weed-tree control in conifer plantings. Manufacturing processes are being constantly improved and the level of TCDD in current 2,4,5-T products is actually much lower than .1 part per million.

In spite of the ultra dilution of this impurity when products carrying it are applied, claims have been made that the use of 2,4,5-T for selective control of weed-trees in forests near Alsea, Oregon resulted in an increase in miscarriages among local women. Based on its study following these claims the U.S. Environmental Protection Agency halted major uses of the herbicide pending a further review of its possible impact on human health. Approvals for a closely related compound known as silvex are also under suspension because it too may contain traces of TCDD.

Defenders of 2,4,5-T point out that many medical authorities involved in a number of separate reviews of these claims question the statistical validity of the epidemiological data on the incidence of miscarriages which formed the basis of the suspension of many uses of 2,4,5-T and silvex. For example, a recent examination of the Alsea, Oregon study* revealed such serious statistical flaws that the reviewers believed no scientific conclusions are possible

Based on a wide range of toxicological investigations there can be no doubt about the acute and chronic toxicity of TCDD to laboratory animals at relatively low doses. Levels in the diet adequate to induce severe toxicity have been shown to also increase the incidence of some types of tumors in rats. Fortunately TCDD appears to be less toxic to humans. Contact with industrial chemicals containing this impurity as a result of improper procedures or accidents in manufacturing operations have occasionally resulted in chloracne, a dermal eruption. Based on many years of experience among chemical factory personnel chloracne, which is believed by medical authorities to provide the first evidence of exposure, may occur without further manifestation of toxicity.

The health status of workers in plants which have manufactured both 2,4,5-T and trichlophenol (the chemical intermediate from which 2,4,5-T and silvex are derived) have been extensively studied. There is no evidence of an increase in the incidence of cancer or other chronic disases even among those who had chloracne at one time or another nor among those who had worked in the plants for as much as 30 years. An accident in a trichlorophenol plant in West Virginia in 1949 resulted in a number of cases of chloracne and a recently published medical evaluation of the 30-year health history of those

---

*A Scientific Critique of the EPA Alsea II Study and Report. Published by Environmental Health Sciences Center, Oregon State University. Corvallis, OR 97331 U.S.A.

involved indicated no long-term effects from this exposure.

The explosion at an industrial plant at Seveso, Italy in 1976 did not involve 2,4,5-T but the same dioxin found as an impurity in that herbicide was released into the atmosphere. Several thousand people were exposed to TCDD at levels tremendously in excess of any possible human exposure associated with the use of herbicides containing this impurity. Seveso and the entire surrounding region has been under intensive medical surveillance since the accident. Three years after the explosion no medical problems associated with TCDD have been identified aside from the skin disorder, chloracne, nearly all of which is reported to have healed.

Toxicological research with monkeys failed to indicate reproductive problems at dosage levels of 2,4,5-T with its dioxin impurity far above likely human exposure, and a three generation study with rats have shown that neither chemical is mutagenic. An EPA investigation revealed no TCDD present in mothers' milk in women who presumably been exposed to 2,4,5-T. Chromosomal investigations at one U.S. manufacturing plant showed no abnormalities in the work force. An area of one square mile at a U.S. air base where aerial equipment was tested during the 1960s received repeated doses of Agent Orange. The sprayed land is now covered with vegetation going through its usual succession of species. Studies with the beach mouse, a major inhabitant of the area, indicated no carcinogenic or teratogenic effects.

TCDD is rapidly degraded by sunlight on leaves, in water and on soil surfaces. Persistence is greater once it is incorporated in soil; however, the compound is quite resistant to leaching. Based on a wide range of analysis of plant and animal life in sprayed areas including the airbase test-area referred to above defenders claim there is little evidence that TCDD persists in the environment in hazardous amounts particularly following label-approved uses. Analytical chemists in Europe and the USA, using recently developed and highly sophisticated instruments have discovered that TCDD is sometimes formed in trace amounts on combustion of organic matter. Thus, doubts are raised that the miniscule quantities sometimes found in the environment were derived from the impurities in herbicides as first assumed.

Although 2,4,5-T is used for weed control in rice in some regions, by far the greatest agricultural application has been for controlling woody brush and other non-nutritious, unpalatable or poisonous plants on range or pasture land (see Chapter 13). Benefits in terms of economic management of woody plant growth along highways and other right-of-ways have been pointed out but according to foresters the greatest benefits to society from 2,4,5-T have resulted from the selective control of weed-trees and the resulting improvement in efficiency of timber and pulpwood production (see Chapter 28).

In their September 1979 report, a statutory Scientific Advisory Panel appointed by EPA to review the question of safety of 2,4,5-T and silvex products strongly favored continued use of these herbicides. The Panel concluded that "residues (2,4,5-T, silvex and TCDD) in water, sediment, aquatic organisms and/or the potential for exposure from herbicide drift do not suggest the possibility of significant risk." The report states, "The monitoring data obtained thus far does not suggest that TCDD derived from commercial 2,4,5-T and silvex exhibits any tendency to accumulate in human food...."*

---

*In December 1980, after a lengthy and exhaustive study, the British Advisory Committee on Pesticides, chaired by the Dean of the Faculty of Medicine at Leicester University, advised that 2,4,5-T can be used safely in the United Kingdom. That government has accepted the committee's conclusions. The summary of their report states:

"There is no valid medical or scientific evidence that 2,4,5-T herbicides harm humans, animals or the environment if they are used in the recommended ways and for the recommended purposes."

At the time of this writing, EPA has taken no action to reactivate suspended uses. "The jury is still out" on the question of safety of 2,4,5-T and silvex products containing traces of TCDD and the outcome of the controversy will depend on further investigations and possibly court actions.*

*Readers interested in the scientific aspects of efforts to resolve this question are referred to: Report of the Scientific Dispute Resolution Conference on 2,4,5-T. Published by the American Farm Bureau Federation, Park Ridge, IL 60068 U.S.A.

# 34

## Pesticide Accidents

Causes of mortality as reported on death certificates have long been the subject of record keeping throughout much of the world and recently authorities in a number of countries have separated accidental poisonings by agricultural chemicals from those involving other substances such as drugs, household products and industrial materials. Although attempts are often made to record intentional poisonings separately from accidents the understandable covering up of homicides and the reluctance of families or friends to admit suicides probably results in some inflation of the accidental category.

Estimates of worldwide mortality related to pesticides have been made by projection from limited data, often from countries that may not be representative. At this time no world figure appears reliable and it seems best to look at each country separately to the extent that meaningful data is available.

Recently a European-based international organization, GI-FAP,* has conducted a country-by-country inquiry into the

*Data on both lethal and nonlethal accidents for countries other than the USA were supplied by Dr. Jacques Cossé, Director General, International Group of National Associations of Pesticide Manufacturers (GIFAP) Avenue Hamoir 12, 1180 Brussels, Belgium. The data was compiled from official government figures, poison center reports and agricultural association records.

incidence of accidental deaths from pesticide exposure. Some governments group fertilizers with pesticides under the heading, agricultural chemicals, so numbers may include mortality of a child who ate fertilizer or a farm worker who used anhydrous ammonia in a grossly careless fashion. The years during which various countries have recorded mortalities associated with agricultural chemicals separately from those resulting from other substances varies considerably but enough data are available in several to give some idea of the frequency of fatal accidents.

In the United Kingdom mortalities related to pesticides are remarkably low considering a population of nearly 60 million. During the three-year period that detailed records have been kept, 1976-78, there was only one death reported among farmers, agricultural workers and the general public. This fine record must be considered as a tribute to the British Pesticides Safety Precautions Scheme as discussed in Chapter 31.

The Scandinavian countries and Switzerland report no deaths for the years records were kept while Austria and the Federal Republic of Germany each averaged about one lethal incident per year. France experienced approximately two and Italy slightly more than three deaths per year from accidents involving agricultural chemicals.

Among the North African countries with enough data to be indicative, Tunisia and Algeria averaged somewhat less than two mortalities annually, however, Egypt reported 38 accidental deaths in 1975, a large proportion of them in homes where children had access to pesticides used to combat household insects. Over a seven-year period of record keeping, South Africa averaged less than four accidental mortalities per year except for 1978 when a collective incident occurred involving the inept preparation of food in used pesticide containers that had not been cleaned.

Turning to the other side of the globe, Australia averaged slightly less than one mortality annually over a six-year period while Indonesia reported an average of 1.3 mortalities per year for the same length of record keeping. Japan with 115 million inhabitants had an average of about 6 accidental deaths per year for the period 1976-78.

Turning to North America, Canada reported one fatal accident for 1975 the only year for which precise data was available but one considered representative. The United States has not had an enviable record for pesticide safety, even considering a population of well over 200 million. Fortunately, as will be seen, considerable progress toward reducing accidents has been achieved during the past two decades. Data on poisonings of all kinds are collected by the National Center for Health Statistics and published annually in *Vital Statistics of the United States*. The nature of accidental deaths is categorized in the preparation of that report, and beginning in 1968 a new category, Accidental Poisoning by Pesticides, Fertilizers or Plant Foods, was established. Suicides and homicides that may have been pesticide related were not included in the above group, but were placed in the category of poisonings of various types that were probably intentional.

The numbers of accidental deaths associated with agricultural chemicals recorded by the National Center for Health Statistics are as follows:

| 1968 ..... 72 | 1972 ..... 38 | 1976 ..... 31 |
|---|---|---|
| 1969 ..... 56 | 1973 ..... 32 | 1977 ..... 34 |
| 1970 ..... 44 | 1974 ..... 35 | 1978 ..... 31 |
| 1971 ..... 43 | 1975 ..... 30 | |

These published records do not detail the nature of the exposure resulting in death but they do break the data down into age categories. As in other countries a considerable proportion of pesticide poisonings involve children which emphasizes the importance of great care in storing materials having potential toxicity. For example, of the 31 lethal incidents in 1978, twelve involved children under ten years of age.

The U.S. Environmental Protection Agency has made estimates of all deaths associated with pesticide exposure (accidental plus intentional) by projection from data recorded by representative hospitals in different regions of the country. Relatively few of the observed episodes in this study were occupationally related. The estimated average number of mortalities attributable to pesticides for 1971 through 1976 (the years for which the statistical projections have been completed) is 53. Considering that this number includes suicides and homicides, while the average of 35 recorded by the Center for Health Statistics for the same years involves only incidents presumed to be accidents, the two sets of data appear in sufficient agreement to justify confidence in their validity.

It was recently pointed out* that an erroneous number, 200 deaths per year related to pesticide exposure and attributed to the same EPA study referred to above, was reported in a paper presented to an American Association for the Advancement of Science Symposium in 1979. Possibly the authors misread a report that gave 200 as the total number of lethal incidents for the three-year period, 1971-73, and mistakenly presented that figure as an annual average. This error has been furthered by quotation of this 200 number in two recent books and is likely to be requoted from them in the future.

A glance at the following table, which presents the total accidental poisoning picture for the years 1976-78 from data collected by the National Center for Health Statistics (the last

*"Illuminating Some Obfuscation About Pesticides," by Keith C. Barrons, *Farm Chemicals*, October, 1980.

three years for which data were available at the time of this writing), will indicate the relative importance of agricultural chemicals as causes of mortality by poisoning.

| Accidental Deaths from Poisoning in the USA — 1976-78 | | | |
|---|---|---|---|
| **Type of Poisoning** | **Number of Deaths** | | |
| | **1976** | **1977** | **1978** |
| Total lethal poisonings | 5730 | 4970 | 6663 |
| Gases and Vapors | 1569 | 1596 | 1737 |
| Drugs and Medicaments | 2839 | 2214 | 3797 |
| Other solids and liquids, excluding agricultural chemicals | 1291 | 1126 | 1098 |
| Agricultural Chemicals (Pesticides, Fertilizers and Plant Foods) | 31 | 34 | 31 |

Neither the fact that of the 6,663 people lethally poisoned in 1978 only 31 died of agricultural chemical exposure (less than one-half of one percent), nor that 104,000 died of accidents of all kinds in that year (including 51,000 in automobile accidents) justifies complacency. The marked reduction in incidence of lethal poisoning from agricultural chemicals in the decade following 1968 is encouraging but gives no comfort to those who may be poisoned through carelessness in the future. How and why did these incidents occur? Will a knowledge of the nature of past accidents help us avoid additional ones in the future?

In an attempt to determine the circumstances surrounding past incidents of pesticide poisonings and to verify the accuracy of death certificates, Dr. Wayland J. Hayes, Jr. of the School of Medicine, Vanderbilt University has, over the last two decades, periodically reviewed records supplied by the National Center for Health Statistics. Initially this critical review was under the auspices of the U.S. Public Health Service. Dr. Hayes and his colleagues had communications with the signers of death certificates where pesticide involvement was indicated. After including cases of questionable postmortem diagnosis, they attributed the following deaths in the U.S. to pesticide exposure during those years the study was made.*

| | | | |
|---|---|---|---|
| 1956 | ..... 152 | 1973 | ..... 61 |
| 1961 | ..... 111 | 1974 | ..... 52 |
| 1969 | ..... 87 | | |

Among the 1973 and 1974 deaths 35 and 27 clear violations of safety precautions were reported. Examples were accidental ingestion of pesticides placed in soft drink bottles, allowing children to play where they could open the valve of a pesticide spray tank, carelessness in playing or sleeping close to grain treated with a fumigant, and failure to keep pesticide containers where children or the less competent could not have access to them. In addition, improper procedures while applying a fumigant and lack of or improper respirators while formulating or mixing the more hazardous pesticide products were reported. There were also a number of pesticide-related suicides and homicides in all years studied while several victims were intoxicated with alcohol when the pesticide ingestion or other exposure occurred. Among the instances of mortality there were five classified as "occupational" in 1973 and seven in 1974, but these were, for the most part, associated with gross carelessness or disregard for elementary safety precautions. No example of death from a pesticide used in the prescribed way was reported.

California physicians report those illnesses of employed persons that are believed to be caused by pesticides and during the six-year period, 1974-79, only one pesticide-related death occurred among farm workers or others who might be exposed while carrying out their jobs. This one occupational death

---

*Mortality from Pesticides in the United States in 1973 and 1974. By Wayland J. Hayes, Jr. and William K. Vaughn. *Toxicology and Applied Pharmacology* 42:235-252 (1977). This paper includes data from those earlier years in which detailed analysis of records was carried out.

which occurred in 1975 involved a structural pest control worker who, according to a California Department of Agriculture report, "handled cyanide in a grossly careless manner." Lethal home accidents or others of a nonoccupational nature in California were included in the data from *Vital Statistics of the United States* presented earlier in this chapter. It has been estimated that California uses 20 percent of all pesticides employed in the USA and because of extensive acreages of fruits and vegetables even a larger proportion of restricted products known to have potential acute hazard to applicators and agricultural workers if carelessly used.

Other important agricultural areas of the USA that use pesticides extensively have had few mortalities related to these chemicals, particularly in farming or other occupations where exposure might occur. In North Dakota, for example, where vast acreages of small grains, potatoes, sugarbeets and sunflower are protected with appropriate chemicals from weeds, insects and fungi, only one mortality was recorded by that state's Department of Health for the years 1977 through 1979. Although this incident occurred in a farm home, it could hardly be classified as occupational as the victim mistakenly drank from an insecticide bottle.

Turning to a representative state in the Corn Belt, an inquiry directed to the Institute of Agricultural Medicine of the University of Iowa revealed that no fatal occupationally related pesticide accidents have occurred in that state during the memory of those on the staff, or since 1965. There have been deaths from home accidents and suicides during that period for which state records were not available. In the cotton-soybean region of the southern USA, the State of Arkansas conducted an intensive study of pesticide-related accidents over a three-year period. The state's physicians reported morbidity and mortality associated with exposure to pesticides to the state Health Department. Their records show a total of four mortalities over the three years, two in the home involving arsenical compounds and two reported as occupational. One of the latter involved a worker who got an undiluted pesticide product on his hands and arms and failed to wash it off, while the other was the pilot of a plane which crashed while applying an insecticide. As his load was dumped before impact, it was not clear that exposure was responsible for the accident.

*Non-lethal Accidents Involving Pesticides.* Of course, mortality is only a part of the picture; we must strive to reduce the occurrence of nonfatal accidents caused by careless exposure. A review of data available from those countries that keep records of agricultural-chemical-related morbidity as well as mortality reveal such a wide range of standards and collection methods that comparisons are of questionable validity. Some numbers merely reflect total calls to a poison control center by a physician or an individual following possible exposure, while in other countries tabulations include complaints where exposure is presumed but no clinical manifestations of poisoning were observed. Some statistics reflect only systemic symptoms while others include cases of skin or eye irritation following pesticide application. Some compilations fail to distinguish between accidents and suicide attempts, the latter an astonishingly common occurrence with pesticides in some parts of the world.

Authorities generally recognize the tendency of people to attribute coincidental headaches or other minor ailments to pesticides, particularly when statistics are being gathered. Most public health officials concerned with poisonings agree that many of the available estimates of the number of pesticide-related minor illnesses sometimes requiring outpatient or non-hospital medical attention are of doubtful validity. In the USA the Environmental Protection Agency states in a report that records of poisoning cases admitted to hospitals "provide the best potential for a reliable data source." With recent stepped-up efforts by health departments of many countries to compile accurate statistics and identify specific risks, a more complete picture of nonlethal accidents should be available in the future. With increased recognition of hazards associated with careless storage and handling of pesticides, the trend should be downward.

Statistics from those countries* that record hospitalizations resulting from pesticide accidents or otherwise classify incidents as serious may be indicative of frequency. In the United Kingdom, with 1.5 million employed in agriculture, 46 cases per year were recorded. This figure includes pesticide manufacturing personnel as well as farm workers and the general public. In continental Europe the following annual incidence of pesticide intoxication requiring hospitalization or otherwise classified as serious has been reported:

>Norway — 10         The Netherlands — 15
>Switzerland — 17

The annual number of pesticide accidents in The Federal Republic of Germany is estimated at 200 based on projections from agricultural liability insurance claims, however the severity of the cases is not indicated.

Tunisia reports 50 nonlethal pesticide poisonings per year over a two-year period and Algeria 33 annually during the ten-year period that records have been kept. South Africa has 56 cases annually classified as serious and Australia reported six people made ill each year by pesticide exposure over the period 1971-1976. Japan reported an average of 184 pesticide accidents annually over the three-year period that records have been kept.

In the USA the Environmental Protection Agency has, since 1971, made estimates of the number of hospital admissions believed to result from pesticide exposure. The data are calculated

---

*Statistics on non-lethal accidents from outside the USA were supplied by GIFAP. See first footnote in this chapter.

by projections from actual admissions to representative hospitals in different sections of the country. That organization's estimates of hospital admissions resulting from accidental pesticide poisonings for the years 1974-1976, the last three years for which their statistical projects have been completed, are presented in the following table:

| Year | Occupational | Non-Occupational (including unclassified) | Total |
|------|------|------|------|
| 1974 | 686 | 1733 | 2419 |
| 1975 | 985 | 1830 | 2815 |
| 1976 | 685 | 1879 | 2564 |

In addition to these estimates of hospital admissions as a result of accidental pesticide exposure, it was estimated that an average of 493 cases of intentional poisoning resulting in hospitalization occurred in each of the three years, 1974-1976. All statistics on pesticide-related mortality and morbidity indicate that children account for a distressingly high proportion of victims. According to EPA estimates nearly 40 percent of hospital admissions in the nonoccupational column in the above table were under five years of age. Let us hope that future statistics will show a safety benefit from the elimination of the more toxic materials from home and garden use and greater care in storage. Let us further hope that with our current drive toward safer application procedures, secure storage, more careful container disposal, and precautions regarding re-entry time, the occupational column will show a downward trend.

Only a few states in the U.S.A. have attempted to collect data on the incidence and severity of illnesses associated with pesticide exposure. During a three-year period, in the mid-1970s, the Arkansas Department of Health sought the cooperation of hospitals and the medical profession in reporting pesticide-related morbitity and mortality occurring within that state.

Of the 129 incidents reported during the three year period, 82 occurred within the home and involved inadvertent ingestion or other exposure to a range of pesticides but predominantly to arsenicals and the rodenticide, warfarin. Seventeen of the 21 cases involving arsenic compounds were hospitalized, and there were two mortalities. The use of arsenicals has greatly declined since these data were compiled. Only a few of those who ingested warfarin were hospitalized and none was seriously ill. Of the entire group of 129 incidents over the three-year period of the study, one-third involved children under 16, largely in the home. Of the 47 occupationally related incidents only three occurred on the farm while 44 involved careless handling in formulating plants, in connection with non-agricultural pest control or other activities. The two occupationally related mortalities during the three-year period of the Arkansas study were discussed earlier.

For several years California has required physicians and hospitals to report to state officials all occupationally related human health incidents believed to have resulted from pesticide exposure, regardless of severity. Cases of localized skin and eye effects have been included in these reports in addition to systemic effects. During the years 1973-1978, the number of persons reported to have a systemic response averaged 587 annually while there were a somewhat greater number of incidents of skin and/or eye irritation.

Only for 1979 are data available from California on the number of persons hospitalized in that state as a result of pesticide exposure in connection with their work. The data for that year as reported by the state's physicians are:

Number of persons reported to have systemic effects 472
Number reported to have eye or skin irritation        545
Number of persons hospitalized                         48
Total days of hospitalization                         137

These limited data from individual states on the numbers of pesticide accidents resulting in hospitalization or otherwise considered serious do not permit extensive comparisons with the EPA projections for the USA as a whole, but the following may suggest need for such an evaluation. Arkansas, with extensive acreage of cotton and soybeans, is believed to use at least 5 percent of the pesticides applied in the USA. With a total of 129 accidents recorded in that state for the three-year study refered to above (43 per year), this would project to 860 annually for the country as a whole compared with the projected average annual figure of 2599 hospitalizations for 1974-76 in the EPA study presented earlier. A similar projection from Arkansas's 16 occupationally related illnesses per year (47 for the three year period of the study) gives an estimate for the U.S. as a whole of 320 compared with EPA's projection from representative hospitals of 784 per year.

If one assumes that the estimate that California uses 20 percent of the pesticides employed in the United States is correct, a projection from that state's 48 persons hospitalized as a result of occupationally related exposure in 1979 gives an estimate of 240 for the country as a whole in contrast to the EPA's projection of 785 annually for the last three years their data has been made available. Even if California accounts for only 15 percent of the total U.S. pesticide consumption, a projection to the country as a whole would indicate 320 hospitalizations, in contrast to the EPA estimate of 785.

Obviously these state data are too meager to justify conclusions, but they do raise a question as to the validity of the EPA projections that should be considered further as hard data from various states become available. With the downward trend in U.S. pesticide accidents as seen earlier in this chapter and the possibility that the recently activated classification of products into restricted and nonrestricted categories will promote greater safety, any future comparisons of various estimates will certainly have to be made for the same period of time.

*Safety is up to us.* Greater pesticide safety can be achieved only when everyone exercises care in storage, application, disposal, and in every facet of use. Considering the known circumstances surrounding pesticide accidents, the following precautions, if universally practiced, would materially reduce their occurrence in the future.

Store all restricted pesticides under lock and key. Others should, as a minimum, be kept where they cannot be reached by children.

Never transfer pesticide concentrates from the original container to another receptacle for storage purposes.

Be sure all empty containers are crushed or a hole put in the bottom so there will be no temptation to use them for water transportation. Then dispose of used containers according to recommendations.

Dispose of unused dilute spray by spreading on ground where people will not contact it. The fastest degradation takes place when there is exposure to sunlight and the action of soil organisms.

Always follow the label with respect to end use, dosage and timing and read and follow all precautions printed on the label.

Wear protective clothing and masks or respirators as specified.

Remove rubber boots and clothing that may be wet by a spray and always wash up appropriately after handling any pesticide. Contaminated shoes should be discarded.

Avoid making applications under windy conditions.

Be sure application equipment is in good order, properly calibrated and not leaking. Do not leave filled spray tanks unattended where children can be exposed in case of a leak or a mischieviously opened valve.

Clean and store application equipment appropriately. Washings from equipment should be run onto ground inaccessible to children, pets and livestock.

Consider the chance that family, neighbors, farm workers, bystanders or passers-by might be contacted as a pesticide is applied. Give warnings or restrict use accordingly.

Observe field re-entry-time recommendations following crop spraying or dusting; for family and all farm workers.

*In Case of Accident.* Recognizing that accidents can happen, industries and many governments provide guidance for those faced with emergencies. In case of ingestion hospitals and physicians can contact national or regional poison control centers for information regarding appropriate antidotes, but the identity of the product, its ingredients, and the manufacturer are essential. It is important to take the container as well as the patient when seeking medical attention for one who is believed to have swallowed a pesticide. Many labels carry the manufacturer's emergency telephone number from which medical personnel can obtain specific toxicology and therapeutic information. In the USA manufacturers provide distributors with Material Safety Data Sheets for each product which give specific information to physicians.

A new emergency telephone source of guidance to the treatment of pesticide exposure, The National Pesticide Telecommunications Network, is available in most of the USA. Physicians, hospitals and poison control centers may call 800-922-0193 from South Carolina where the project is based or 800-845-7633 from the other contiguous 47 states. This phone is manned by clinical pharmacologists 24 hours a day, but they need information on the known or suspected pesticide that may have been ingested.

Immediate first aid should be provided in the event of eye

or skin contact with pesticide concentrates. For eyes, irrigate immediately with fresh water for five minutes. Body surfaces contacted should be washed off in flowing water or a shower. Remove all contaminated clothing and wash before reuse. Use extra care with shoes. In case of inhalation remove to fresh air and consult medical personnel if the exposed person shows an effect.

Spills of pesticide concentrates at home or on the farm are best handled by absorbing liquids or diluting powders with soil, sand or some other noncombustible absorbent and then scooping into a container. Modest quantities can then best be disposed of by spreading on unused land where animals or children will not contact it and where it is unlikely to be carried off by rain. The degrading effect of ultraviolet light, soil microorganisms, and exposure to the elements are such that modest quantities of pesticides are more safely disposed of in this way than to place in a dump or down a drain. For safe disposition of larger quantities of pesticides taken up with an absorbent material, consult the manufacturer.

In the USA the National Agricultural Chemicals Association has a Pesticide Safety Team Network ready to assist police and firemen in minimizing the risk of injury from significant accidental spills. To obtain help from a regional team, call 800-424-9300 or for those outside the 48 contiguous states call collect, 202-483-7616. The Chemical Transportation Emergency Center (CHEMTREC), operated by the Chemical Manufacturing Association, provides these phones with 24 hour answering service. CHEMTREC then relays information to appropriate teams, who in communication with the manufacturer of the product involved can be dispatched promptly to assist in cleanup. Direct contact with the manufacturer whose emergency phone number is on rail cars and packages is also advised. A key precaution in the event of major leaks, spills or fires is to keep the public away. Liquids should be diked to prevent spread and runoff into streams.

# TRADE-OFFS

In writing the preceding pages I have tried to play the role of a reporter and deal with what I believe to be facts, or are perceived to be facts by others. In discussing controversial issues, for example Chapters 32 and 33, an attempt was made to present both sides of the pesticide coin as viewed by different people. If my personal opinions showed through here and there without identification as such, I apologize for departing from ethical newspage journalism. In this discussion of trade-offs I will shed my reporter's role and express some personal views. In journalistic parlance the following paragraphs will serve as my editorial page.

## Pesticides' Hidden Benefits

In considering pesticide trade-offs—risks versus benefits—the obvious plusses of assured food supply and health protection are well recognized. But there are also hidden benefits, ones little known to the public and ignored by anti-pesticide advocates. Examples are the contributions of chemicals to energy conservation and the protection of soil and other natural resources.

Energy requirements per unit of food production generally go down as yield of crops and efficiency of livestock and poultry feed conversion go up, at least to a point of diminishing returns on energy inputs seldom reached in the world of practical farming. Control of pests not only contributes to a

223

higher average level of productivity, but also makes investment in other inputs less risky and more likely to be optimized. Can you imagine a vegetable grower gambling the expensive labor required for establishing a crop and the fertilizer needed to bring it to harvest if he did not have those fungicides and insecticides often required to protect his investment? Or can you conceive of a poultry raiser investing in the prodigious quantities of feed consumed each day in a modern broiler operation without the assurance that coccidiosis would be kept at a low level by one of the anticoccidial drugs routinely employed? Certainly, more energy would be required to produce each serving of chicken or vegetable if effective pest control chemicals were not available.

In recent years, agriculture's spectacular increase in productivity per unit area of land has greatly lessened the acreage that would otherwise have been planted to annual crops. Much of the land thus unfarmed would have been hilly and highly subject to erosion. Today it is largely in permanent pastures and forests. Pest control chemicals have thus played a positive role in soil conservation through their contribution to yield increases. An example is the remarkable enhancement of potato yields, a considerable segment of which resulted from better insect and fungus control. According to statistical data kept by the United States Department of Agriculture, the average per acre production of this crop was less than 70 hundredweight of marketable tubers during the 1930s, but more than 240 hundredweight in the decade of the 1970s. Similar increases in potato yields have occurred in most countries heavily dependent on the crop.

Herbicides have provided valuable additional contributions to soil conservation by making farming possible with greatly reduced tillage. Wind and water erosion, as well as the root-inhibiting soil compaction that results from excessive use of tractors, have often been minimized. The new no-till methods being evaluated for agronomic and environmental adaptability in many areas of the world can reduce soil erosion tremendously, wherever there is need to grow crops on sloping land.

Herbicides have made a further contribution to soil conservation by encouraging vigorous forage through the elimination of competing, non-nutritious, unpalatable, and sometimes poisonous plants. Soil erosion losses from a well-managed grass or grass-legume pasture have been shown to be smaller than from one infested with woody scrub. More efficient production of forest trees is still another land- and energy-saving benefit of modern herbicides.

Nature has ordained that there shall be voracious insects and organisms of decay ready to attack most organic substances. Stored food has the added threat of hungry rodents. The losses of stored products that would be sustained without the contribution of chemicals to integrated protection would place a further drain on the world's land and energy resources. One has only to observe the destruction of grain by rodents and insects in the developing world to appreciate the conservation role of pesticides. Wood preservatives, antimildew agents for paint, bacterial inhibitors for cutting oils, and slime control agents for cooling towers are examples of specialized conservation chemicals now widely employed in the technically advanced world. Energy, soil, products of farm and forest—all must be conserved with diligence, and pest control chemicals are making a significant contribution.

## Balancing Safety Margins

The saccharin episode in the United States, following so quickly upon the banning of cyclamates, served to emphasize that many organic substances, both natural and synthetic, can be shown to be carcinogenic in one or another laboratory animal if administered at doses greatly in excess of likely human intake. The fact that there is no evidence of an increase in human cancer resulting from either of these sweetening agents after years of use should cause even the most wary to reflect on the wisdom of the Delaney Clause. As presently interpreted, this provision in U.S. regulatory law prohibits the addition to food of any substance that induces tumors in laboratory animals at any dosage, no matter how high. Valuable uses of at least ten pest control chemicals have been banned since passage of the Delaney Clause, and numerous promising materials have never been heard of beyond research facilities, where they fell by the wayside when toxicological invstigations—as required by regulatory agencies—showed them to be probable carcinogens at doses greatly in excess of likely human exposure.

There is certainly reason to question the practical meaning of carcinogenicity tests with a single species, particularly tumor-prone strains of mice. Likewise there is reason to question the net benefit to man of preventing the use of chemicals that help assure an adequate food supply only because grossly exaggerated doses have a tumorigenic effect on an inbred strain of laboratory animal. In my view, the current inflexible interpretation of the Delaney Clause, with no allowance for degrees of possible risk or the benefits on the other side of the ledger, must be modified or reinterpreted.

Members of the United States Congress are considering legislation that would introduce the concept of degrees-of-risk, as well as a consideration of benefits, into regulatory procedures. Representative James G. Martin of North Carolina has proposed an independent panel of scientists which would rule whether a given chemical presents a risk of cancer or other illness if used in the proposed manner, this decision to be followed by the weighing of risks versus benefits by regulatory agencies.

Although Congressional action to clarify the law's intent and to provide guidelines for its future administration might be the preferred way to balance risks and benefits, many think such action is unlikely. Another approach would be a reinterpretation by regulatory officials of the word *appropriate* found in the Delaney Clause, which reads:

No additive shall be deemed to be safe if it is found to induce cancer when ingested by man or animal, or if it is
found, after tests which are *appropriate* for the evaluation of
the safety of food additives, to induce cancer in man or animal . . . (italics mine).

Researchers conducting tests for possible carcinogenicity
customarily run the dosage range as high as the animal will tol-
erate without death resulting. Hence, the dosages employed are
often far in excess of likely human exposure. With compounds
having a relatively low acute effect, the highest nonlethal doses
may place a severe burden on the animal's elimination system,
including the liver. It is believed by some that the resulting
pathological manifestations are often in reality a response to
this overloading of the system, rather than true carcinogenicity.
What is so sacred about these traditional ways of conducting
such experiments? Why not a protocol for carcinogenicity
tests, with the highest level at an agreed-upon multiple of anticipated average human intake, for example 100 to 1? Would not
such a margin for safety be quite *appropriate*?

To me, it is highly *inappropriate* to provide for an infinite
margin of safety against the possibility of a miniscule increase
in the incidence on cancer, via the Delaney Clause, but at the
same time ignore the need for a reasonable margin of safety for
our food supply. When DDT was banned in some countries
a decade ago, we had just such a margin of safety for crop
protection in the form of alternate chemicals. Now, with outlawing of one pesticide use after another, that margin has evap-
orated, as has much of the incentive for industry to continue research aimed at safer and more effective crop and livestock
protection materials.

As promising as some nonchemical methods of pest control
may be, those who promote the idea that they alone can defend us against insects, fungi, nematodes, etc. without the aid
and the backup of pesticides are proposing that we all play

Russian roulette with life's number one essential: our food supply. A more flexible interpretation of the Delaney Clause, based on tests at an agreed-upon multiple of likely human intake, would go a long way toward assuring food for the future.

## Some Hidden Benefits to Health

How easy it is to give lip service to the concept of risk/benefit ratios as a basis for regulatory decisions, but how difficult to adequately measure both the numerator and denominator. Health risks from pesticides, real and problematical, have been abundantly investigated and reported, but there has been far too little attention paid to health benefits.

Some environmental benefits of good pest control are quite subtle: conservation of energy and other resources; reduced soil erosion; less land required for crop production with more available for alternate uses. Likewise, benefits to health are often hidden from our view. Aside from their well-recognized role in vector control to minimize malaria and many other diseases, pesticides make a positive contribution to good health by helping to assure an adequate and constant supply of a wide variety of foods.

Far more than tons of grain or other basic foods are involved. Few days go by without the media extolling the virtues of abundant fruits and vegetables in our diet. We like these protective foods, but so do many pests: fungi, mites, insects, nematodes, and others. If fruits and vegetables are to be produced in volume, good pest management practices are imperative. A number of advances in genetic, biological, and cultural controls are helping growers raise these crops more efficiently, but nonchemical methods alone are seldom adequate. Pesticides are still vital, if we are to have a volume of these health-giving natural foods sufficient for all.

In spite of increased recognition of the importance of fruits and vegetables to good health, supplies are not keeping up with demand. One important negative factor is the supply and cost of manpower for these labor-intensive crops, but equally important, I am convinced, is the unavailability of a range of pest control chemicals. So many pesticides have been banned in the name of health protection that growers have become reluctant to expand acreage of fruits and vegetables and, indeed, some have shifted to corn or other crops where the loss of a chemical "tool" is less likely to spell economic disaster. With their high labor costs, many growers cannot afford the risk of possible severe losses due to pests.

Will we in the United States continue the absurdity of eliminating the last vestige of carcinogenic risk—via the Delaney Clause of our food and drug law—only to jeopardize the good health nutritionists unanimously agree is promoted by an abundance of fruits and vegetables in the diet? Many pesticide uses have been removed from approval labels only because it is believed, by some, that the miniscule traces present in one or another crop may slightly increase the incidence of cancer, often projected in the range of one occurrence in one million or more individuals. Assuming that this does occur, which many toxicologists think unlikely, should we not, even so, balance this risk to an occasional individual against the certain blow to good health for millions as a result of fewer fruits and vegetables in their diet?

In her book *Preventing Cancer*,* Elizabeth Whelan attributes 30 to 35 percent of cancers to "imprudent diet." One of her recommendations for a diet less likely to encourage this dread disease is, "Eat more fruits and vegetables." Even if one discounts her estimate of 30 to 35 percent by a considerable

*Elizabeth Whelan, *Preventing Cancer* (W. W. Norton Co., New York, 1978).

degree, should we not question the relative impact on health of:

1) *A trace of a compound found to be carcinogenic only at exaggerated doses in laboratory animals, the likely human exposure of which many authorities consider below a threshold level,* compared to
2) *The loss of production "tools" certain to result in a reduced supply of health-giving fruits and vegetables?*

Let's carefully consider these trade-offs before banning additional crop protection chemicals.

## The Antipesticide Lobby

Man's errors tend to spawn over-reaction, and the sometimes ill-advised use of pesticides in the past has brought forth abundant over-reaction in the form of a veritable antipesticide lobby. Its supporters at times disclaim the target of eliminating all pest control chemicals, and they may give lip service to the value of pesticides as a method of "last resort" for integrated control programs, but their relentless campaign to discredit chemical control methods at every opportunity reveals their true goal.

The supporters of this antipesticide crusade tend to be grossly unobjective, even though their pronouncements are often veiled with the cloak of science. Every instance of confirmed or hypothetical hazard is overblown, frequently out of all proportion, and often with very questionable interpretations of data. They seldom appear to look for—let alone see—the benefit side of the picture. The word trade-offs is not in their vocabulary.

I was introduced to the blind unobjectivity of the antipesticide people several years ago while making a presentation to a Gordon Research Conference, in which I mentioned progress in control of nitrification through the use of a fertilizer additive. As I finished my talk, up jumped a well-known environmentalist to condemn any "tampering" with the natural process of nitrification. I countered with a second explanation of the localized and temporary nature of the inhibition of the soil's nitrifying bacteria and the degradation of the additive to simple chemicals found in the world around us. Then, I re-emphasized that, aside from helping farmers produce better crops when heavy rains cause leaching, the inhibitor saves energy through more efficient fertilizer utilization and, at the same time, reduces the chances of nitrate pollution of water and nitrous oxide contamination of the atmosphere.

All to no avail; his mind was made up without an examination of either risks or benefits. This knee-jerk reaction to the very thought of a chemical contributing to the management of a pest is characteristic of the far-out environmentalists who make up the antipesticide lobby.

An example of recent antipesticide propaganda is Daniel Zwerdling's article entitled "The Pesticide Plague," which appeared in the Washington *Post* for March 5, 1978. His general conclusion that pest control chemicals are worthless, based on an assortment of half-truths and untruths, is representative of the current "party line" of the lobby. Particularly disturbing was his condemnation of the new no-till methods of farming made possible by some remarkable herbicides. Like my antagonist at the Gordon Research Conference, Zwerdling apparently made no effort to comprehend data on benefits before broadcasting his blanket condemnation. No-till farming is a new practice being evaluated around the world to identify its advantages and disadvantages and to determine just where it fits. No one expects the method to replace tillage in all situations, but on some soil, it is now well established that no-till methods greatly reduce soil erosion—still our greatest single source of water pollution. Further, this new cropping system often saves energy. Zwerdling catalogued every objection that has been raised in these widespread trials, not once mentioning the agronomic, environmental, and economic benefits that have been proven for many areas.

Antipesticide propagandists can be pretty careless with numbers, if doing so makes chemicals look bad. An example is the myth that the United States suffers 200 accidental deaths per year from pesticide exposure. Table 101 of a publication of the U.S. Environmental Protection Agency* gives the number 200 as the estimated *total* lethal incidents for the *three years* of the study, 1971-73. But a recent book,** referring to the same report, states: "The total estimated mortality from pesticides is about 200 per year." In his book *The Politics of Cancer,**** Samuel Epstein also refers to "approximately 200 people estimated by EPA to die annually from pesticide poisoning..." Was one author quoting the other, or did both misread the table in the EPA publication? Regardless, this is hardly the kind of scholarship expected of university scientists, presumably dedicated to the truth.

These authors must have had the annual publication *Vital Statistics of the United States* in their university libraries— which would have given them the same data presented in Chapter 33 showing, since 1970 an average of 36 accidental pesticide mortalities associated with agricultural chemicals. They even could have obtained more recent data from the Environmental Protection Agency, which, combined with the data in Table 101 of the report referred to above, would have given them an average of 54 lethal incidents per year, including suicides and homicides—a far cry from the 200 number in their books

---

*National study of Hospital Admitted Pesticide Poisonings 1971-1973, Office of Pesticide Programs, U.S. Environmental Protection Agency.

**David Pimentel and John H. Perkins, eds., *Pest Control: Cultural and Environmental Aspects* (Boulder, Colorado: Westview Press, 1980).

***Samuel S. Epstein, *The Politics of Cancer* (San Francisco: Sierra Club Books, 1978).

published in 1978 and 1980.

A common trait of the antipesticide people is to relate pounds of chemicals used to health and environmental hazard—the "pesticides are bad per se" misconception. Of course, with undesirable materials, more pounds could mean greater risks, but with many of today's pest control products, safety to both health and the environment can be enhanced as volume of usage goes up. The nitrification inhibitors and also herbicides for reducing erosion-promoting tillage are cases in point. Wood preservatives and the pesticides needed for integrated stored grain protection are further examples where the world needs more, not fewer pounds.

## Understatement and Overstatement

The pest control story is usually quite incomplete when told by the antipesticide lobby or the segments of the media that lobby appears to dominate. Hyperbole is a part of their approach, and a balanced picture is not one they choose to paint.

A recent NOVA program on U.S. Public Television laid great stress on cotton insect problems in the Rio Grande Valley of the United States and Mexico. There is little doubt that there were sometimes, in that area, unwise uses of insecticides, which exacerbated pest problems through the destruction of predators, nor is there doubt that the resulting insect problems there were a factor in greatly reduced acreage of cotton, but NOVA might well have left a listener having little knowledge of North American agriculture with the impression that cotton production everywhere was in extreme difficulty because of uncontrollable pests and that there would soon be little of this fiber available. Indeed, cotton farmers everywhere have serious insect management problems but, according to Department of Agriculture data, the U.S. grower produced 476 pounds per acre in the 1970s, compared with only 235 pounds per acre in the period 1930 through 1949—the two decades before modern insecticides became available. Of course, there are other factors which contributed to improved yields, but the implication that the American farmer has done nothing right in managing his sometimes overwhelming cotton insect problems ignores the facts indicated by these marked yield increases. For the most part, he has done the best he could with the knowledge and tools he had at his command, and a doubling of yields indicates that he has not done too badly. Now, with more scientific monitoring techniques and more precise knowledge of life cycles and the impact of various pests at different stages of cotton plant devlopment, the producer will be able to do a more precise job of insect management with reduced likelihood of a repeat of the problems of the Rio Grande Valley.

The true cause of health and environmental protection is being done a disservice by recent misleading overpublicity regarding integrated pest management, often referred to as IPM. Indeed, considerable progress has been made in devising superior pest control systems, and fewer pounds of pesticides may sometimes be required. But seldom are the biological, cultural, and genetic phases of such systems adequate by themselves. Contrary to impressions left by publicity of the antipesticide lobby, a chemical, more often than not, plays a critical role in successful IPM systems. Certainly pesticides are not merely a "last resort" segment, as often implied.

Actually, proven IPM methods ready for farmers to use with confidence have been worked out for relatively few pest situations. Further research aimed at successful and reliable systems should certainly have continuing public support, but the importance of improved pesticides to fill their role in IPM must not be overlooked. Unfortunately, antipesticide propaganda and sluggishly administered regulatory programs are discouraging industrial research aimed at the safer and more selective pesticides that will be needed if IPM systems are to realize their full potential.

Propaganda by the antipesticide lobby is not difficult to

spot: little, if any recognition of benefits, the implication that nonchemical methods can do the whole job, and a "last resort" attitude if there is any admission that pesticides may sometimes be needed. Name calling is a part of the game. I've been identified as a paranoid member of the agri-business mafia, while government agricultural authorities and university specialists are often depicted as part of a conspiracy to foster more pesticidal chemicals on an unsuspecting public. As for the people who spend their hard-earned money on chemical tools to protect their crops and livestock, their health, and their worldly goods, they are by implication likely to be depicted as stupid—too stupid, that is, to accept every inadequately field-tested nonchemical method being proposed.

Of course, there are some zealous pesticide salesmen striving to make their business go, but remember that they and their employers have given us the tools we all depend on for a part of our defense. Practically all new chemical pest management tools have originated in industrial laboratories. Possibly one can find a public agricultural specialist or a university professor "in bed" with the chemical industry, but during nearly half a century of professional employment on "both sides of the fence"— with two agri-businesses and on the faculty of two universities—I never ran into one. As for the stupidity of farmers, well, there are only about half as many of them in the industrialized world today as there were thirty years ago. It has been a rough business to survive in, and I doubt there are many stupid ones among those who are still around, providing us with our daily bread.

## Understanding Chemicals

Unfortunately, the media seldom have occasion to use the word *chemical* except in connection with a transportation accident, a manufacturer's culpability in the disposal of wastes or the carcinogenic potential of one or another substance we have found useful. It is not surprising that the myth has evolved that chemicals are bad per se. Understandably, people with limited scientific knowledge fail to recognize that all substances, including living matter, are made up of chemicals. They often make fervent believers in the myth that natural materials are safe, while synthetics are inherently hazardous. Reporting of real and alleged risks, without a look at the other side of the coin, seldom contributes to public understanding.

Do not all with a scientific background share responsibility for helping explain the chemical facts of life to the public? Do we not have a responsibility to aid the media in distinguishing between hazards due to carelessness or ineptitude and the lack of appreciable risk from present-day pesticides when used according to label instructions and precautions? Once the safety of a pesticide used in an approved manner is brought in focus, do we not have an obligation to help explain its benefits, and the costs of not using the approved pesticide material? Certainly the environmental activists with scientific understanding should join in an attempt to present a balanced picture of the role of chemicals in our life.

One need not go back to the vicious anti-DDT campaign of a decade ago to find instances of one-sided reporting, a mere cataloguing of problematical as well as confirmed risks associated with the current "whipping boy" among industrial or agricultural chemicals, often without so much as a mention of the usefulness to mankind of the substance in question. Some publicity lacking in balance may result from unawareness on the part of the reporter, often fed only the risk side of the equation by environmentalists seeking support for their preconceived position. Some may stem from editors' and publishers' insistence on sensationalism, while still more comes from the environmentalists themselves, anxious to continue riding a good "publicity horse" en route to research grants, self-aggrandizement, and even political power.

I am reminded of a series of stories in a Florida newspaper

regarding the adverse environmental impact of phosphate mining—an operation all agree needs modifying to assure minimum water pollution and the grading of disturbed land back into a usable condition for forests or pastures. The articles could well have left the impression, with an uninformed reader, that phosphate rock was dug from the ground by greedy industrialists solely for the joy of messing up the landscape! How can the public help make wise trade-off decisions, often resulting in regulation, if people know nothing of the importance of phosphates in formulating mineral supplements for efficient livestock feeding? Or if they are unaware of the role of phosphate fertilizers in enabling farmers to produce abundant crops on soils that nature did not endow with enough of this vital element? How can they know about the many industrial uses of phosphorous, including manufacture of the organophosphate insecticides (now widely used in lieu of the organochlorines, because of their shorter residual life and resulting environmental safety)?

In this age when we are so dependent on chemicals, yet have such an inadequate grasp of their attributes, pesticides have often been among the least understood and most controversial. There have been many contributors to the current polarization of thought and actions relative to pest management, a situation that ill serves the cause of food abundance, good health, and conservation of precious resources. It will take the action of many to banish this polarization: the media, farmers, environmentalists, pesticide manufacturers, and researchers seeking improved methods of pest control. You too, can play a part.

If we are to have the benefit of a range of chemicals as a part of integrated pest management programs, regulations must take not only the negative, but also the positive side of the matter into consideration. However, regulations that reflect sensible trade-offs of risks versus benefits are unlikely to come about unless an informed public begins to demand balance on the part of legislators and regulators. Agricultural organizations and the scientific community concerned with defending mankind against the world's myriad of pests are striving to bring about such an understanding, but they need help.

You can help. You can vote for sensible trade-offs by making your views known at every opportunity—to your friends, to the public, and particularly to your legislative representatives.

## *Yes, Pesticides Are Really Necessary!*

That, in a nutshell, is my answer to those with environmental concerns who have been encouraged to believe that we can defend ourselves against the world's innumerable pests by non-chemical methods alone. Unless we prefer a return to raw nature, defenses by one means or another are imperative. Much progress has been made in encouraging natural controls and in devising various nonchemical methods, but integrated systems involving these alone seldom provide adequate protection. Chemicals are a part of most successful integrated pest management programs, and for many pests, where only modest progress has been made in devising cultural, biological or genetic controls, pesticides are still the first line of defense.

Improved monitoring methods are enabling us to employ chemicals at the time they will make the greatest contribution to pest management—often an early application to suppress a first "brood" or to reduce populations when the host is most vulnerable to the pest. The "pesticides as a last resort" philosophy, the notion that the wisest use of a chemical is to wait until an infestation is great and serious losses imminent, is often unsound both economically and environmentally, as well as from the standpoint of health protection and natural resource conservation.

Research will undoubtedly give us plants with greater resistance to various pests, as well as new approaches to biological control, but within the vista of my crystal ball, chemicals will continue to form an integral part of most improved systems of pest management. I also perceive the possibility that more effective, more selective, and safer pesticides can be discovered if that research goal is vigorously pursued. But the picture beyond the possibility stage is far from clear. Will regulatory policies and public attitudes encourage the research needed to bring about such discoveries, and will they encourage the investment needed to make technical advances available to the public? Only with free enterprise incentives and without the disincentives of erratic and sluggishly administered regulatory programs—as experienced in recent years—will we reap the benefit of these possibilities. Without the streamlining of controls, many safer, more specific, and more effective pesticides already discovered —but not yet commercialized—will never benefit mankind.

# Index